THE BOOK OF
FLIGHT TESTS

ALAN BRAMSON
FRAeS

Past Chairman of the Panel of Examiners
Liveryman of the Guild of Air Pilots and Navigators

Arco Publishing, Inc.
New York

For James Gilbert in England, Hugh Whittington in Canada, Tom Chalmers in South Africa, Macarthur Job in Australia, Dennis Shattuck in the United States and other editors who have encouraged me to write airtest reports for their excellent aviation magazines.

Published 1984 by Arco Publishing Inc.
215 Park Avenue South, New York, N.Y. 10003

First published in the United Kingdom in 1984 by Martin Dunitz Ltd, London.

Library of Congress Cataloging in Publication Data

Bramson, Alan Ellesmere.
  The book of flight tests.
  Includes index.
  1. Airplanes—Flight testing.   I. Title.
TL671.7.B73   1984   629.134'53   83–25751
ISBN 0–668–06152–9

Phototypeset by Input Typesetting Ltd, London
Printed in Italy

CONTENTS

ACKNOWLEDGEMENTS

For allowing me to fly the aircraft described in the following pages I should like to thank the manufacturers mentioned and their agents. I am particularly grateful to the enthusiastic team at Martin Dunitz for the very high standards of production they have achieved at a time when so many publishers are lowering their sights. Finally my thanks to Barry Newman for his splendid prints of my photographs.
Alan Bramson, 1984

The publishers would like to thank the following individuals and organizations for their permission to reproduce illustrations: Aerospatiale, pages 17 and 88; Air Portraits, page 94; Avions Marcel Dassault, page 216; Avions Pierre Robin, pages 34–5; Beech Aircraft Corporation, pages 101–3, 118 (below) and 189; British Aerospace, pages 165, 198–9, 219, 220, 223; Cessna Aircraft Corporation, pages 24, 50–1 (above), 77, 132, 152 (below), 157, 174–5 and 210; De Havilland Aircraft of Canada Ltd, page 179 (below); Eagle Aircraft Services Ltd, pages 111, 140–3 and 186–8; Embraer, page 194; James Gilbert, pages 29, 38, 44–5, 47, 54, 60–1 (below), 79, 109 and 122–3; Pilatus Britten-Norman Ltd, pages 160–1, 169; Piper Aircraft Corporation, pages 53, 56–7, 99, 114 (below), 128–9, 138, 148 and 184–5; Short Brothers Ltd, page 203.

All photographs not listed above were taken by the author.

ABBREVIATIONS

AC	Alternating Current
ADF	Automatic Direction Finder
ADI	Attitude Director Indicator
agl	above ground level
APU	Auxiliary Power Unit
ASI	Air Speed Indicator
AVGAS	aviation petrol (gasoline)
AVTUR	aviation turbine fuel
BTU	British Thermal Unit
CAA	Civil Aviation Authority (UK)
DC	Direct Current
DME	Distance Measuring Equipment
eshp	equivalent shaft horsepower
FAA	Federal Aviation Administration (USA)
ft	feet
HF	High Frequency
hp	horsepower
HSI	Horizontal Situation Indicator
IAS	Indicated Air Speed
IFR	Instrument Flight Rules
ILS	Instrument Landing System
ISA	International Standard Atmosphere
ITT	Interstage Turbine Temperature
kg	kilogramme
km/h	kilometres per hour
kt	knots (nautical mph)
kw	kilowatt
m	metres
mpg	miles per gallon
mph	miles per hour

NAV/COM	Navigation/Communications
nm	nautical miles
OAT	Outside Air Temperature
psi	pounds per square inch
rpm	revolutions per minute
shp	shaft horsepower
STOL	Short Take-off and Landing
TAS	True Air Speed
TBO	Time Between Overhauls
TMA	Terminal Manoeuvring Area
VFR	Visual Flight Rules
VHF	Very High Frequency
VOR	VHF Omni-directional Radio Range
VSI	Vertical Speed Indicator
VTOL	Vertical Take-off and Landing

V-code

V_1	Decision speed
V_2	Take-off safety speed
V_{at}	Target threshold speed
V_{mc}	Minimum control speed
V_{mca}	Minimum control speed, air
V_{mcg}	Minimum control speed, ground
V_{mcl}	Minimum control speed, landing
V_{ne}	Never exceed speed
V_{no}	Normal operating speed
V_r	Rotate speed
V_y	Best rate of climb speed
V_{yse}	Engine-out best rate of climb speed

PREFACE

Why write a book of this kind? Well, for a good many years now, I have been flying all manner of light and business aircraft in the course of preparing articles for a number of aviation magazines; in fact nine of them throughout the world. Types flown range from diminutive single-seat racers to business jets and thirty-seat turboprops.

The idea that perhaps one day I might write a book of airtests based on this experience did cross my mind from time to time but the prospect took a turn towards reality when a few years ago I was invited to give a colour-slide illustrated talk at a big seminar run by the Limerick Flying Club in Eire. Pilots came from all over Ireland, north and south, and the lecture, in which I described what it was like to fly some of the aircraft I had flight tested, was well received. This book is an extension of that idea, and the forty planes I have chosen to feature between its covers represent as wide a spectrum as possible of the current general and commercial aviation scene.

'How can you cope with flying such widely differing aircraft types?' I am often asked. Most pilots will agree that the more types you fly the easier it is to feel at ease in a strange aircraft. Contrary to what you might imagine, big aircraft are not more difficult to handle than little ones, often quite the reverse. Taking it to extremes, the little Formula One racers, which do more than 220 mph (354 km/h) on only 100 hp, are twitchy, capricious and demand a fine touch. Conversely, large aircraft are more stable and predictable, the main problem being to understand the various systems, which can be very complicated

Viewing light aircraft over the past forty-five years, and I am referring to single-engine trainers and tourers – remarkably little progress has been made in terms of performance. For example, in 1936 the Percival Vega Gull would fly four adults in a reasonably comfortable cabin and cruise at 140 kt. Rate of climb and range was better than a Piper Arrow, which enjoys the advantages of retractable landing gear and a more modern engine weighing 180 lb (82 kg) less than the old 200-hp Gipsy Six. Likewise, the Miles brothers, Fred and George, were manufacturing light planes in the mid-1930s that, in terms of performance and, in some cases, comfort, compare more than favourably with modern designs.

To be fair, some features of light singles have improved, brakes and engine starting in particular. And of course it is only in more recent times that lightweight, easy-to-use radio has become available; there was no such thing as VOR before the war and ADF was a big, heavy luxury, reserved for the airliners of the day. The big improvements have come in the form of safe light multis, piston or turboprop powered, while the business jet could not have been imagined forty years ago.

These days aircraft designers have learned most of the tricks of avoiding really unpleasant handling characteristics, while manufacturers are closely controlled in so far as they must conform with clearly defined airworthiness requirements. So in terms of safety and predictable handling there should be no dark corners.

However, such matters as safety and straightforward handling are minimum requirements; there are other aspects of aircraft that are not covered by the airworthiness regulations yet are of vital importance to the pilot or prospective owner. Is the aircraft comfortable? Is the cabin wide enough or do you sit so close that you almost feel like 'consenting adults', to use the legal term? Can you see out of the aircraft (the Americans are often bad at this)? Is it possible to talk without having to shriek into your neighbour's ear? If you fill the tanks, can you carry more than just a pilot and his toothbrush, and does the plane offer an honest cruising speed for its power?

Many of these questions may be answered in black or white terms. Payload/range and all other performances are expressed in figures. Likewise it is easy enough to measure cabin width or quote maximum baggage capacity. It is when you try to enter the realms of handling, design appeal, cockpit layout and so forth that the aviation writer is on less sure ground because in these areas you are dealing with matters of taste – and to quote the old saying: 'One man's meat is another man's poison.' Handling is an interesting aspect to write about, because while most professional pilots, civil or military, will climb into, say, eight or ten different aircraft and probably give the same verdict on which is excellent, which is average and which is unmentionable, the amateur of low experience both in flying hours and types flown is often nonplussed when he hears the pundits comment, 'what dreadful ailerons' or, 'it flies like a bus.'

So when reading the various test reports it should be borne in mind that the figures quoted are fact, but the likes and dislikes are mine. I like to believe they are motivated by the experience of flying many aircraft types rather than by prejudice.

A.E.B. 1984

1. HOW TO USE THIS BOOK

For easy reference the airtests have been assembled in six chapters: Single-Engine Trainers; Single-Engine Fixed-Undercarriage Tourers; Single-Engine Retractable Tourers; Piston-Engine Twins; Turboprops; and Jets. Each aircraft is described and reported on in the same format and, whenever possible, measured performance figures are quoted as well as those claimed by the manufacturers, the latter appearing in a Facts and Figures data panel at the end of each test.

First, a word of warning about interpreting the figures claimed by the manufacturers. If you read their brochures they will tell you their pride and joy will lift, say, four passengers with 200 lb (91 kg) of baggage, cruise at 170 kt and fly a nonstop 1,000 nm with enough fuel for a 45-minute hold. What the manufacturers do not make clear is that it cannot achieve all these performances at the same time. Thus Brand X, which may have a 500-nm range with full tanks, will only fly 140 nm when all seats are occupied and maximum baggage is carried.

To get nearer to reality, you have to consult the flight manual; but it does not end there, because the empty weight quoted for most light aircraft, even some of the twins, takes no account of the essential equipment that must be added before they are fit to fly. Often the instruments included in the empty weight are inadequate, even for flight in good weather. The Americans started this unhealthy fashion of quoting unrealistic empty weights, and now, in a mood of 'if we can't beat 'em, join 'em', other firms in different parts of the world have followed suit. Perhaps if the manufacturers would agree at least to quote empty weights for an aircraft equipped with a full flight panel, but excluding radio (that is a matter of personal choice), we might be looking at more meaningful figures. But as things now stand, by the time a NAV/COM installation has been fitted, along with an instrument panel, anti-collision beacon, navigation, landing and instrument lights, pitot heater and such extras as wheel fairings (which can significantly influence both climb and cruise performance), you must add 80–100 lb (36–45 kg) to the published empty weight. When full airways radio is required and perhaps a simple autopilot that figure can easily reach 130 lb (59 kg). Naturally, an increase in empty weight is accompanied by a corresponding decrease in useful load. Remember that this extra 130 lb is equivalent to an average-size woman or 18 Imp/21.6 US gal (82 litres) of fuel.

It is not easy to understand the policy adopted by the larger general aviation manufacturers. For example, Piper quote an empty weight for the Warrior (see Chapter 3) that does not even include a turn and balance indicator or basic lighting for night flying (although four ashtrays are considered important enough for inclusion in the basic price). The same applies to their Seminole light twin (see Chapter 5). You are even faced with adding over 110 lb (50 kg) in essential equipment to the Piper Seneca (also featured in Chapter 5). Yet if we move up the scale a little to the more powerful Aerostar (Chapter 5), lighting and flight instruments are included in the empty weight, but not avionics.

The moral to all this is clear: read the small print and guard against being taken in by manufacturers' brochures. I am not suggesting that they deliberately set out to fool the prospective buyer. It is just that they do not give you all the facts. In principle, the bigger, more powerful aircraft are marketed with complete airways capabilities, but that is of little consolation to the would-be owner of a light tourer who must ensure that the aircraft he is considering will fly his family over a long enough range for his needs. In the airtest reports I quote typical equipped empty weight and useful loads.

Presentation

The test reports are presented in the following format:

- **Background** A brief historical account of the aircraft's development.
- **Engineering and design features** A description of the hardware, systems and aerodynamic characteristics.
- **Cabin and flight deck** A rundown on the passenger area and the pilot's 'office'.
- **In the air** A blow-by-blow account of what the aircraft is like to fly – warts and all.
- **Capabilities** What it can and, more important, what it cannot do.
- **Verdict** An assessment of how it rates vis-à-vis similar aircraft.
- **Facts and figures** Basic details of dimensions, weights, performance, and so on.

At the end of each chapter there is a star-rating table for at-a-glance reference that compares the important features of all the aircraft in each category.

Understanding the numbers

For a test report to be of value the figures quoted must mean something to the reader. The following sections should help explain their significance.

Aircraft weight

The maximum authorized take-off weight takes into account such factors as structural strength, take-off performance, stalling speed and rate of climb. In the case of twin-engine aircraft the limiting factor is usually the need to provide an acceptable single-engine climb-out in the event of engine failure during or soon after take-off.

It has become the fashion to quote a ramp weight. This is slightly higher than the take-off weight because it allows fuel for engine starting, taxiing and the power checks. In the case of a light twin such as the Piper Seneca III (see Chapter 5) we are only talking about an extra 23 lb (10 kg) of fuel but in a big aircraft the ramp weight allowance would represent significant additional payload for the hard-pressed operator.

From the ramp weight (or when this is not quoted, the maximum authorized take-off weight), we subtract the equipped empty weight (not the meaningless standard empty weight quoted in the manufacturers' brochures) to arrive at the all-important useful load. How this is used is at the discretion of the pilot. He could fill the fuel tanks and probably have to restrict the number of passengers or their baggage. Alternatively the cabin could be filled, but usually this would entail a reduction in the amount of fuel that can be carried.

Some aircraft, usually larger designs, have a maximum landing weight. This is understandable in the case of a large passenger jet, where 65 tons of fuel may be burned off at the end of a long flight, but a landing weight restriction is, in my view, a bad design limitation when it applies to a light aircraft. One small twin must burn off 237 lb (108 kg) of fuel – the equivalent of 33 Imp/40 US gal (151 litres) – and that would mean flying around for eighty minutes before a landing could be made, had take-off occurred at maximum weight.

Finally, some of the larger aircraft have a maximum zero fuel weight. This is to stop pilots creating excessive bending loads in the main spars by loading the cabin from floor to ceiling, while carrying little fuel in the wings. Old aircraft had no such restriction, but they were built like a suspension bridge, and useful load suffered as a result. These days the aim of the game is to spread the load along the wing span, so if you take the zero fuel weight and subtract the equipped empty weight: that will tell you how much can be put in the cabin. The rest of the useful load must go either in the wings, if it is fuel, or in the engine nacelle baggage lockers (when these are provided).

Performance

Cruising speeds vary according to flight level, aircraft weight, temperature and power setting. In the case of single-engine, piston-powered aircraft without turbochargers it is most common to quote a 75 per cent power fast-cruise setting, usually at around 8,000 ft where the throttle will be wide open to attain that power. The other figure most often mentioned is the 65 per cent power long-range cruise (usually at around 12,000 ft), but some of the aircraft flight manuals will provide tables of figures against power settings, flight levels and temperatures. Cessna flight manuals are particularly good in this respect. Turboprops and jets cruise at high power settings because turbines are then at their most efficient. Usually they recommend a high-speed cruise and a long-range cruise.

Range Wind component apart, range is influenced by aircraft weight, temperature, power setting and flight level. The flight manual will provide suitable tables or range graphs, and with light aircraft it is usual to quote 'with-reserve' ranges which allow for start-up, taxi to the runway, power checks, take-off, climb to cruising level, the flight itself, descent, landing and taxiing to the apron. There is also provision for a forty-five-minute hold or diversion, but this should be checked because some manufacturers cheat a little by computing the forty-five minutes at a very reduced power setting.

Although the aircraft brochure usually highlights the 75 per cent (fast) cruise, range is almost invariably quoted for the 65 per cent power setting, no doubt to enhance the capabilities of their product. The data tables in this book give both 75 per cent and 65 per cent power cruising speed whenever possible.

Rate of climb is quoted at sea level and under standard temperature conditions.

Take-off and landing roll is of limited value. More important is:

Take-off and landing distance, since in the case of aircraft of up to 12,500 lb (5,700 kg) weight these assume that a 50-ft screen or barrier must be crossed during the initial climb and on final approach. So take-off distance is to the point where the aircraft reaches 50 ft above ground level (agl) and landing distance is measured from where the aircraft has descended to 50 ft agl.

Landing and take-off performance is affected by aircraft weight, temperature, airfield elevation, type of surface, incline, decline or level surface and wind strength. Most flight manuals provide tables giving ground rolls and distances to 50 ft in the case of take-off (or from 50 ft while landing) against the variables I have listed.

For simplicity, landing and take-off performance figures quoted in the Facts and Figures tables at the end of each test relate to a hard runway at sea level, under standard temperature conditions and zero wind. Most turboprops and a few business jets are fitted with reverse thrust and when a landing distance entails using this facility it will be mentioned.

Engine power
In the case of piston engines this is quoted in good, old-fashioned horsepower (hp) since this remains the most widely used unit.

Turbopropeller engines (turboprops) are slightly more complicated, because in addition to the power delivered to the propeller, known as shaft horsepower (shp), some residual thrust is ejected from the final turbine stage(s) and since this adds to total propulsion, an equivalent shaft horsepower (eshp) is sometimes quoted. To give some figures, the Garrett TPE 331-9 develops 865 shp plus 105 lb (48 kg) of thrust resulting in a total output of 907 eshp.

Jet engines are power-rated in terms of static thrust measured in pounds. The noisy and thirsty straight turbojet has almost entirely been replaced by the fanjet in which most of the gas energy is used to drive a large-diameter fan. This in turn generates a tube of relatively slow-moving cold air which encircles the hot and excessively vocal gas jet from the turbine stages, greatly reducing noise levels and significantly improving fuel economy. I am sometimes asked how to convert the thrust of a jet engine into horsepower and I am afraid the short answer is: not easily. As a matter of interest a British Aerospace 125/700 business jet (see Chapter 7) cruising at 30,000 ft and 460 kt would, at average weights, be developing the equivalent of 3,275 hp.

Engine life
Most engines have a 'life' expressed in terms of running hours or time between overhauls (TBO). Small piston engines run for 2,000 hours, while larger and more complex units have a life of 1,500–1,800 hours. It is sometimes possible to get an extension when the engine is in particularly good condition.

Turboprop engines have proved to be more reliable than piston units. This is hardly surprising, considering the crude nature of 'suck-squeeze-bang-blow' motors, with their red-hot valves snapping shut while pistons flash up

and down, frantically peddling around their tortured crankshaft. When you realize that at 2,500 rpm a piston flashes up its cylinder, stops, flashes down again, then stops (a cycle repeated more than forty times a second), it is nothing short of a miracle that these sons of the age of steam manage to function at all, let alone with a fair degree of reliability.

All turbine engines, prop or fanjet, rotate smoothly in one direction, there is no stop-and-reverse of heavy moving parts and vibration is almost nil as a result. Typical TBOs for turboprop engines range between 3,000 and 6,000 hours, although there is a move towards replacing the 'life' system with one called 'on condition'. As the term implies, the engine is inspected at regular intervals and allowed to run forever so long as parts showing evidence of distress are replaced.

These days fanjets are usually operated 'on condition', although prior to the adoption of this very sensible procedure a TBO of 8,000 hours was not unknown.

All the performance figures I have given assume an aircraft in good condition, at maximum weight and equipped as recommended by the manufacturers. It is unreasonable to expect the claimed cruising speed from a fixed undercarriage aircraft if, for your own reasons, the wheel fairings have been removed. Under these circumstances some aircraft will give away up to 10 kt, at the same time showing a marked decline in climbing performance.

Understanding the terminology
I want to conclude this introductory chapter by explaining some of the terms that may be unfamiliar to inexperienced pilots or those new to aviation. It is important that these are understood, as they appear repeatedly in the airtests.

The airframe

Flush rivet When drag must be kept to a minimum the metal is either pressed in around the rivet hole or it is countersunk to allow a rivet head to be formed in smooth conformity with the aircraft surface. Obviously, flush riveting is more time-consuming than ordinary snap-head riveting and it is therefore more expensive.

An alternative method of attaching metal skins to frames and stringers is with the tubular-shaped pop rivet, which is formed by pulling through a mandril (which looks like a large pin with a tapered head) with a special tool. Nowadays it is fashionable to eliminate rivets in larger aircraft in favour of Redux bonding, a form of adhesive which must be cured in a large oven called an autoclave – surely a better method than drilling lots of little holes and filling them with metal again.

Light alloy Aluminium is too soft a metal for practical use but when small amounts of copper, manganese and other elements are added, an alloy called duralumin results, which is more than four times stronger than aluminium. However, it is less resistant to surface corrosion, so aircraft sheeting usually takes the form of duralumin with very thin surfaces of pure aluminium hot-rolled on each side.

Fibreglass The resin-filled, glass-cloth boat or car body is well known and for some years now such unstressed aircraft parts as engine cowlings, wing-tips and wheel fairings have been made in this material. However, in West Germany great progress has been made in the building of all-fibreglass airframes. At first, only sailplanes were constructed in this way but a new breed of two-seat, motorized, fibreglass gliders has emerged, and the sheer perfection of finish on these aircraft has to be seen to be appreciated (see pages 40-1). Carbon fibre spars are being used on some and several manufacturers see a growing future for this, at present, very expensive material.

Anticorrosive treatment In some parts of the world corrosion is not a problem. For example, even the cheapest cars last forever in South Africa. However, in northern Europe, the northern United States and Canada, moisture, salt atmosphere and the effects of industrial pollution all accelerate the corrosion process.

Some manufacturers offer corrosion proofing as an optional extra. This is a bad arrangement because often the treatment is sprayed on after the aircraft is riveted together and it is then too late to protect the metal interfaces where airframe rot is most likely to occur.

A well-constructed airframe should be chromate treated (or the equivalent) *before* the individual parts are riveted together. Larger, more expensive aircraft are 'wet' assembled, that is, additional anticorrosive sealants are applied to the adjoining faces immediately prior to riveting.

Fatigue Most metals, some more than others, will eventually crack when subjected to reverse-load flexing over a long period. This is a particularly important consideration with pressurized aircraft. To cater for the risk of catastrophic structural failure as a result of cracking, all aircraft are now designed on a 'fail-safe' basis, with other parts arranged to take over from the one that has failed (appropriately named 'multiple-load' stressing). Crack retarders or stoppers are often incorporated in the design.

Integral fuel tanks When part of the actual wing structure is made fuel-tight and used to carry AVGAS (aviation petrol) or AVTUR (aviation turbine fuel) it is said to have integral tanks. The alternative is separate light-alloy or flexible tanks.

Autoslats These leading-edge devices are mounted on over-centre linkages. At high angles of attack their centre of pressure moves the slats off the mainplanes to form slots through which high-pressure air from below the wings is able to flow. This high-pressure air is directed by the slats to move over the wing-top surfaces and thus delay the stall.

Buffet inducers Sometimes called root spoilers, these small strips are often fitted to the wing leading edge as it nears the fuselage. At high angles of attack the airflow over the inner parts of the wing is made turbulent by the strip. This causes a buffet to be felt on the control wheel as the disturbed air flows back over the tail surfaces, thus providing a natural stall warning. At still higher angles of attack the inner portion of the wing stalls prematurely, and because the outer areas are unstalled there is little tendency for a wing to drop at the stall.

Washout Another method of preventing a wing drop at the stall is to reduce the angle of incidence at the wing-tips (angle of incidence being measured between the fuselage longitudinal datum and the wing chord line). This is called washout and it ensures that the inner portions of the wing are always flying at a larger angle of attack than the tip areas. Consequently the inboard areas stall first while the tips remain flying.

Frise ailerons So called because they were designed by the late Leslie Frise of the Bristol Aeroplane Co in 1919, these ailerons have remained in constant worldwide use ever since. Their purpose is to reduce or eliminate the symptoms of aileron drag, and they are attached to the wing by inset hinges. When the aileron goes down, as it would on the outside of the turn, it remains in smooth conformity with the wing. Simultaneously the up-going aileron protrudes its leading edge below the undersurface of the other wing, causing additional drag and since this is on the inside of the turn the adverse yaw effects of aileron drag are eliminated.

Simple flaps Plain, hinged surfaces which are not particularly effective at producing additional lift or drag.

Slotted flaps These open up a slot between the flap leading edge and the wing and provide a 65 per cent increase in lift.

Split flaps These are now considered old-fashioned but are, in my view, better than slotted flaps because they provide almost as much additional lift along with a bigger drag increase – useful during the landing approach.

Fowler flaps These have the additional talent of increasing the wing area by first moving back from the trailing edge on tracking before depressing. Normally these flaps provide a 90 per cent lift increase, but larger aircraft have multiple-slotted Fowlers which open like a Venetian blind to produce more than a 100 per cent increase in lift.

Speed brakes Jet aircraft cruise at very high speeds that are sometimes near their limiting Mach number. And a heavy plane moving at, say, 400–500 kt takes some slowing down when the time comes to join with other traffic. To assist in the task, power-operated speed brakes are fitted which may be used at any speed. Usually they take the form of wing fences.

Lift-dump Most jets and some turboprops have this facility which is often achieved by depressing the flaps to a large angle. The aim is to reduce lift after landing so that full advantage can be taken of the aircraft's brakes on what might be a wet, slippery runway.

Powered nosewheel steering Larger and heavier aircraft are fitted with powered nosewheel steering. Usually the nose strut is hydraulically actuated and controlled by the pilot through a nosewheel steering knob positioned for use with the left hand. There are variations to this arrangement and these are described in the various airtests.

Antiskid brakes Various automatic antilocking devices are fitted to the brakes of larger high-performance aircraft, a typical example being the Maxaret system.

The engine
Most piston engine terms will be well known to the reader so this section is confined to turbine units.

Power lever This relates to turboprop engines. Throughout most of its movement a power lever behaves like the throttle of a piston engine. But it may be lifted and brought back behind its idle stop to select Beta mode.

Beta mode When the power lever is moved back past the idle stop the propeller constant speed unit is inoperative and blade angle is then under the mechanical control of the pilot. In Beta the blades are at or near an angle where no thrust is developed. This is useful while taxiing, since most turboprop aircraft suffer from excess idling thrust and in consequence tend to taxi as though the last one at the runway is a cissy. Further backward movement of the power lever selects:

Reverse thrust With the power lever brought back behind the Beta mode position, blade angle takes on a reverse pitch and engine power increases to provide reverse thrust for aerodynamic braking during the landing roll. The flight manual usually recommends discontinuing reverse thrust at speeds below 40 kt. This is to avoid the risk of blade erosion due to small stones and other debris being thrown up from the surface.

Single-shaft engine This type of turboprop engine has a single shaft on which the turbine stages are mounted to drive the compressor stages. A suitable reduction gear is needed for the final drive to the propeller. Small turbines can turn at a staggering 50,000 rpm.

Two-shaft engine These are sometimes called free-turbine engines but in essence this type of turboprop is in two main parts. The prime mover is a set of compressor stages which is driven by one or more high-pressure turbines, the assembly being known as the gas generator or core. The high-energy gas is directed to the second part of the engine – one or more low-pressure turbines which drive the propeller through a separate shaft via suitable reduction gears. Such engines are started and shut down with the propeller feathered (to stop rotation of the propeller with the minimum of delay).

Some free-turbine engines have a condition lever in addition to the power lever and the propeller lever.

Condition lever In the rearmost position this cuts off fuel to the engine. During the starting sequence it is moved to the GROUND-IDLE position, and prior to take-off the fully forward FLIGHT position is selected to provide the correct fuel flow for take-off, climb and cruise. These days the trend is to eliminate the condition lever and incorporate the fuel cut-off with the propeller lever so feathering the blades automatically starves the engine of fuel.

Torque limiter A lot of power is developed by the turbines and at low altitudes it would be very easy to overload the propeller transmission. Some turboprop engines are provided with an automatic torque limiter which is of particular value during the take-off at a time when pilots have better things to occupy their attention than little engine dials.

Temperature limiter At higher cruising levels turboprop engines can overheat to the point where very considerable (and expensive) damage results. To guard against over-cooking, some engines are provided with an automatic temperature limiter.

Engine instruments (turboprops)
Unfortunately the manufacturers have yet to standardize the units shown on the following turboprop engine readouts.

Torque meter This measures the amount of power being supplied by the power turbine stages to the propeller. It may be calibrated in lb:ft, psi or, best of all, percentages of maximum. There will be a red line, above which 'thou shalt not venture'.

Interstage turbine temperature (ITT) This is usually calibrated in degrees Centigrade with a red line which must not be exceeded. Apart from safety purposes, the ITT is useful for fine power adjustments.

Rpm indicator In a single-shaft engine there is only one of these, but a free-turbine unit has two, one for the gas generator and the other reading propeller rpm. In some aircraft the latter is calibrated in old-fashioned rpm, but there is a growing tendency to show all readings as percentages of maximum.

Flow meter Quite separate from the fuel contents gauges are the flow meters, one for each engine, which indicate how much fuel is being burned in pounds per hour.

Engine instruments (fanjets)
Fanjet engines have all the readouts described for turboprops, with the exception of torque meters. With these engines there is an rpm indicator for the fan (labelled N_1) and another giving the speed of the gas generator N_2).

The systems

Pressurization The average passenger is comfortable flying at pressure altitudes of 8,000 ft or so. Above 10,000 ft it is advisable for pilots to use oxygen; and if full benefit is to be derived from turbocharged piston engines, turbo-props or jets, which are at their best cruising in the 20,000–40,000-ft region, oxygen is essential and so is cabin pressurization.

In the case of turbocharged piston-powered aircraft, cabin pressure is provided by the engine turbocharger units. Their capacity is very limited and it is a feature of piston-powered aircraft that the throttles should not be closed while flying at high altitudes, otherwise a rapid drop in cabin pressure could cause severe discomfort and possible ear damage to people suffering from catarrh.

Turboprop- and jet-powered aircraft obtain cabin pressure from the high-capacity compressor stages which enable comfortable conditions to be maintained in the cabin even with the power at the idling setting.

Pressurization controls vary but they usually entail setting the planned cruising level and climb rate. An instrument is provided to indicate cabin altitude, which typically would not exceed 8,000–10,000 ft when the aircraft is at its maximum cruising level.

An automatic 'squat' valve on one leg of the undercarriage is fitted to ensure that all cabin pressure is dumped on landing. Otherwise, exit from the aircraft would be rapid and entertaining when the door was opened.

Hydraulics The hydraulic pressure that retracts and lowers the landing gear may be provided by engine-driven pumps or a self-contained electro-hydraulic power pack. Sometimes the flaps are also actuated by this system and, in more complex aircraft, so are the nosewheel steering, speed brake (for jets) and power-assisted wheel brakes.

Electric system Most small piston-engine aircraft are fitted with alternators but turboprops and jets have DC starters which convert to DC generators when the engine is running. Consequently, AC inverters must be provided for some instrument and radio equipment. Sometimes the flaps, undercarriage and nosewheel steering are electrically operated.

Ice protection If an aircraft is to be used for serious transport, ice protection is essential. The windscreen and propeller(s) may be electrically heated and the flying surfaces (wings, tailplane and fin) will need either pneumatic de-icing boots or anti-icing strips. Rubber leading-edge boots are cheaper to install in the first instance but there is a performance penalty when they are in use and the pilot must exercise skill if they are to be really effective. The other disadvantage is their limited life (two to three years according to the amount and type of flying). The replacement cost can be higher than that of the original installation. Personally I prefer metal foil anti-icing strips through which de-icing fluid is sprayed when the need arises. There is no performance penalty, the system has a long life and it can be fully automatic, even on a light aircraft.

Engines usually have their own in-built ice protection.

The instruments

Attitude director indicator (ADI) With more and more instruments threatening to take over the cockpit and exclude the crew, not only have designers had to overcome the problem of finding panel space, they have also been faced with the difficulty of pilot scan – the ability to take in readings from so many different sources. Consequently instruments were amalgamated or integrated. Thus we have what used to be called an artificial horizon tied in with some form of VOR/ILS command bar or needles which tell the pilot to FLY UP, FLY RIGHT, FLY LEFT, and so on, during an instrument approach.

Horizontal situation indicator (HSI) Occupying the space below the ADI, where you would find the direction indicator in a light trainer (directional gyro in the United States), is another integrated instrument which combines the readouts of a gyro compass with those of a VOR. The deviation or LEFT/RIGHT needle rotates with the compass card and a model aircraft etched in plan form on the centre of the instrument glass creates the illusion of flying towards, away from or along the required radial.

The ADI and the HSI are more commonly found on the larger piston twins, turboprops and jets but the HSI is by no means uncommon in the more well-equipped single-engine tourers and it is only a question of time before the ADI is as commonplace in this class of aircraft as the artificial horizon.

True airspeed indicator There are relatively inexpensive versions of these instruments on the market. A small knob allows you to set temperature against height in a little window. This moves the ASI scale around the instrument face so that the needle reads TAS.

My main criticism is that most of these instruments are too difficult to set because the figures are very small. However, some excellent instruments are now available that automatically compensate for height and temperature.

Vertical readouts Some of the small business jets are fitted with engine instruments that look for all the world like vertical thermometers. Many pilots have reservations about these, particularly since of necessity, electronic digital readouts are built into the instruments to allow fine adjustment.

The avionics

Most readers will be acquainted with much of the VOR, ADF, ILS, DME and communications equipment that is now commonplace, even on single-engine aircraft. Less familiar may be some of the following 'black boxes' which, at present, are normally only fitted to turboprops and jets.

Single sideband HF In some parts of the world, VHF communications are of little value due to the terrain or the distances involved. Under normal conditions single sideband HF can provide excellent communications to any part of the world. However, it is relatively heavy to install and very expensive.

Omega/VLF These very low-frequency navigation systems are yet to reach their ultimate level of reliability but they are nevertheless in quite widespread use. Omega uses eight dedicated transmitters situated throughout the world while VLF takes advantage of eight rather more powerful NATO transmitters. These are not dedicated to aerial navigation and are therefore not guaranteed to be available twenty-four hours a day, year in, year out.

Both systems are similar and you can buy aircraft equipment capable of being switched from one set of transmitters to the other. In either case only three stations need be received and the control unit is very similar to that used for Inertial. The pilot inserts the lat and long of each point to be flown over (called waypoints) and the equipment will proceed to indicate heading to steer, ground speed, wind velocity, time to next waypoint, lat and long of present position (usually to an accuracy of plus or minus 2 miles (3 km) anywhere in the world), heading back to base if a return is necessary – you name it, it does it!

Omega/VLF is about 25 per cent the weight of Inertial and very much cheaper to install, and not surprisingly is becoming popular with corporate aircraft operators. This is real area navigation as opposed to those much cheaper (and less accurate) courseline or vector computers.

Colour weather radar   The ability to depict cloud density according to colour is an obvious advantage and it is interesting to note that colour weather radar has proved more reliable in service than the old monochrome equipment.

Another valuable talent of the more advanced colour weather radar set is its ability to display check-lists for pre-take-off, emergency procedures, and so forth.

Altitude alerters   When an aircraft is capable of reaching, say, 30,000 ft in less than twenty-five minutes and it may be required for air traffic reasons to descend at a high rate, an altitude alerter is essential. This takes the form of a small dial which may be set to the required altitude/flight level. As the aircraft climbs or descends towards that level a warning bell will sound in advance. It will also sound if the pilot departs by plus or minus 200 ft from the level set on the dial. At the same time the command bar in the attitude director indicator will give an appropriate FLY UP or FLY DOWN indication. Altitude alerters are now commonplace on light turboprops and business jets.

By now you should be well armed to understand fully the airtests that follow – and to join me in the light planes, turboprops and business jets that are featured in the main part of this book.

2. SINGLE-ENGINE TRAINERS

I once discussed the problems of designing a light single-engine trainer with an eminent aeronautical engineer, now retired, who took a leading part in producing the revolutionary Vulcan bomber in Britain. It was one of the first delta-wing bombers to enter service with any airforce and more than twenty years after its first delivery to the British Royal Air Force it remained a remarkable technical achievement and a great aircraft. Yet this same engineer who had taken such a leading role in the Vulcan project confided that many years previously he had become involved in designing a low-powered, two-seat trainer, and that it was one of the most difficult tasks of his distinguished career.

Now this may come as a surprise to many readers but let us examine what is required of a good trainer.

- It must be strong enough to endure rough handling at a flying school and higher-than-average utilization (in light-plane terms).
- Maintenance must be easy.
- Spares must be readily available and cheap to buy.
- Visibility must be good. This is particularly important while training within a busy airfield circuit. High-wing monoplanes are particularly blind towards the inside of a turn and I regard that as a serious disadvantage.
- The aircraft must be easy to enter and leave. With some designs the occupants have to go into training preparation for the task.
- Whether or not seating is tandem or side by side (and there are certain advantages to the more social sit-beside-me arrangement for civil training) the cabin must be large enough to provide a reasonable degree of comfort.
- Noise levels must be kept to a minimum. Most light trainers make very poor classrooms because they are cramped and for most of the time there is so much engine and slipstream distraction that the student only hears one word in ten of what is being said by the instructor.

- There must be adequate panel space for all the flight instruments and, when instrument training is contemplated, an airways radio fitted.
- Performance must be good enough for meaningful training. Rate of climb should allow the instructor and student to reach a safe height for spinning/aerobatics without wasting too much time (700 ft/min at least). Cruising speed should be high enough to allow sensible navigation exercises. Range should be sufficient to encourage club members to hire the aircraft for interesting trips.
- A good baggage area is needed if club members are to hire the aircraft for trips abroad.
- The aircraft should be cleared for spinning to meet the requirement of those states where spin training is mandatory (and quite right too, in my opinion).
- Aerobatic (acrobatic in the United States) capabilities would be an advantage to broaden the number of training roles open to the club or school.
- The aircraft should handle well and be safe but not too easy to fly accurately.
- Engine power should be the lowest possible in order to attain good fuel economy but not to the point where all-round performance suffers.
- The plane should be good to look at – customer appeal is important.
- The ability to tow a glider, while not essential, could be an advantage at some schools although usually at least 160 hp is required for that task.
- Our wonderplane must be cheap.

Now clearly this list is a pretty tall order and, in my view, there have been few truly great elementary trainers since the era of Tiger Moths, Chipmunks and Stearmans. But those days are long since past and we have to face reality. Five of the trainers in this section are in widespread use. The last described is less well known but it is an interesting new development.

The flying classroom. Aerospatiale Rallye cabin.

Piper PA-38-112 Tomahawk

The light trainer designed by committee

Background

For many years the American Piper Co built and sold in considerable numbers the dear old Cherokee, a '2 plus 2' trainer which was built like a battleship but something of a blancmange to fly. Then in 1976 they announced that a new trainer was being designed on the basis of a questionnaire sent to no fewer than 10,000 American flying instructors. Now that was a dangerous and courageous thing to do because in Britain such a questionnaire would have brought forth 10,000 different answers, perhaps even more. The most interesting result of the questionnaire was that although for many years the American FAA has not required spin training for a United States private pilots' licence the majority of those gallant 10,000 said that they wanted spin capability built into the new aircraft. In my view they (and Piper) are right about spinning and the FAA are wrong, but that is another story.

One of the earliest Tomahawks to depart the shores of America – in fact the seventh to be built – arrived in a box at CSE Ltd, Oxford Airport, England and soon afterwards I was able to fly it for a number of magazines. At the time I commented that, here and there, Piper's new bird looked more than a little delicate and predicted that it would probably not stand up to life at a busy school. Now, four years later, I have to tell you I was wrong. Apart from being one of the most important Piper merchants outside the United States, CSE operate possibly the largest flying school in the world and, I am reliably informed by those I trust, their Tomahawks have stood up to utilizations of more than 1,000 hours per annum without trouble. But then I should not be surprised; the seemingly obvious is rarely the case in aviation. CSE's Tomahawks, by the way, have 5,500 hours of trouble-free flying on MOGAS to their credit.

Engineering and design features

The Tomahawk is based upon an untapered, high-aspect-ratio wing which uses a modern GAW-1 airfoil section. Rather crude, slab ailerons are attached by a piano-wire hinge along their top surface, and plain flaps of similar construction hinge down to reveal a gap between their top surface and the wing. Two fuel tanks with a total usable capacity of 25 Imp/30 US gal (114 litres) are located forward of the main spar.

The fuselage is slender, with the cabin area raised to allow good visibility through 360 degrees with few

obstructions. A strong hoop, braced by a pair of large-diameter tubes, is constructed behind the door to protect the occupants in the event of an undignified arrival terminating in the inverted position. Metal doors in the engine cowling are released by those excellent catches that were a feature of the Cherokee, while stays keep them open to allow maintenance down to rocker box level.

Perched high up on the swept fin, in fact almost at the top, is the tailplane with its separate elevators. You will look in vain for a trim tab; 'hands-off' flight is achieved by kind assistance of spring tension as the pilot adjusts his trim control (the old Moths had that arrangement in the 1920s). The rudder has a metal tab which is adjusted (bent) on the ground to provide balanced cruising flight.

Simple, spring steel legs support the mainwheels, and the nosewheel steers through the rudder pedals. Two fixed steps assist entry via the wing walkways. A roof handle must be turned before individual handles will release the two small car-type doors.

Cabin and flight deck
The cabin is about 43 in (109 cm) wide at shoulder height – which is 7 in (18 cm) wider than the Cessna 152 (see the next test). The otherwise good all-round visibility is spoiled to some extent by rather thick door pillars.

Fuses are situated low down on the right of the instrument panel, and the engine controls, consisting of a throttle and a mixture lever, are on a neat little quadrant with a fuel selector between them that points to the contents gauge of the tank in use – idiot proof, which is as it should be.

Carburettor heat is selected on a lever which moves in a slot cut out of the instrument panel. Standard brakes take the form of a Cherokee-type lever with a lock-on button but individual toe pedals can be specified as an option. A small lever between the seats places the flaps in three positions: UP, 21 degrees and 34 degrees, which is the maximum setting. A good-quality primer is fitted, and there is plenty of room for a full flight panel and all the radio one would normally fit in this type of aircraft. Electric switches are in a row along the bottom of the left-hand panel, and each pilot has an adjustable ventilator.

The seats provide good support; as they adjust forward for shorter legs so they rise slightly to cater for proportionally lower eye levels.

As fitted to the Tomahawk, the Lycoming 0-235-L2C engine develops 112 hp. It started readily using the key operated ignition/starter switch. Nosewheel steering is very positive and the aircraft is easy to taxi.

In the air
At near maximum weight I lined up on the runway, opened the throttle and timed 12 seconds to lift-off. There was no tendency to swing, and the initial rate of climb was timed at a satisfactory 750 ft/min. For this flight we were about 50 lb (23 kg) below maximum weight so Piper's claimed 718 ft/min for a fully loaded Tomahawk sounds reasonable.

A placard giving power settings at various heights is conveniently fixed to the roof and at 3,000 ft I set the throttle at 2,450 rpm to attain 75 per cent. That gave us an IAS of 95 kt which trued out at 99 kt; but if you climb to 7,100 ft the 75 per cent power setting will cruise a maximum weight Tomahawk at 108 kt. However, for

OPPOSITE: Piper Tomahawk. The cabin enclosure provides good visibility in most directions.

Instrument and control layout is logical and convenient. Note the foolproof fuel selector which points to the tank in use.

training, 65 per cent power would be more economical and quieter for the loss of 8 kt cruising speed. Noise levels are acceptable, but nothing to write home about.

A trainer should reproduce the real aviation world, not some cotton wool, protective version of it, so I was glad to see that at the stall there was sufficient wing drop to be convincing without being frightening. Flaps-up we stalled at 52 kt IAS, flap 21 degrees took off 4 kt and full-flap brought the stalling speed back to a modest 45 kt.

To spin the aircraft you close the throttle, hold up the nose slightly and at about 60 kt apply full rudder in the required spin direction. Then the wheel should be held fully back. Entry is smooth and the aircraft settles into a slowish rate of rotation in a steep nose-down attitude.

There is a fair amount of twanging from the rear fuselage during a spin as flexing metal gives its impersonation of a Jamaican steel band. Recovery, using the standard procedure, first results in a momentary increase in spin rate, then rotation ceases, the rudder is centralized and the aircraft may be eased out of the dive. I feel that in training terms the Tomahawk has an excellent spin, provided pilots recognize that for a modern light plane the nose attitude is steep. This should not discourage easing the control wheel forward during the recovery. Incidentally, Piper recommend a nonstandard spin recovery in their flight manual. We could all do without this kind of thing. An inexperienced pilot finding himself in an uninvited spin should not have to ask himself 'do I ease the wheel forward or firewall it in one go'. Standard spin recovery in an emergency saves lives: Mr Piper, please take note.

There is slight lateral stability, powerful yaw damping and the aircraft is stable in pitch. The controls, if not harmonized in the classic proportions are nevertheless pleasant with light, moderately effective ailerons. Among modern American light trainers the Tomahawk is better than average in terms of handling and visibility.

Initial approach at 65 kt aiming for 60 kt 'over the hedge' works nicely and in common with most modern light planes the Tomahawk is easy to land.

Capabilities

Maximum take-off weight is 1,670 lb (758 kg). Standard empty weight — that useless figure I castigated in Chapter 1 — is quoted as 1,109 lb (503 kg). By the time you have added such essential items as flight instruments, night lighting, NAV/COM and a VOR indicator the equipped empty weight is 1,156 lb (524 kg), leaving a useful load of 514 lb (233 kg). If you put two 170-lb (77-kg) adults in the aircraft along with the 100-lb (45-kg) baggage allowance (which is stowed in a luggage area behind the seats) 74 lb (34 kg) is left for fuel — that is, 10 Imp/12 US gal (45 litres). And that would provide a range of about 180 nm at 108 kt (75 per cent power), or 188 nm if you elect to use 65 per cent power and settle for an 8-kt decrease in cruising speed.

Full tanks will provide with-reserve ranges of 452 nm

Tomahawk in the air.

around to be fastened by a row of rivets on each side of the fuselage. Furthermore, the doors have a catch which is held in place by the top handle; these catches are made of bent, heavy-gauge wire but wire nevertheless. So as I said at the beginning of this test report, my first impression was that the Tomahawk would not withstand life at a busy flying school. Well I am happy to say that events have proved me wrong – the aircraft is tougher than it looks. Structural problems with the fin where it joins the fuselage are now being overcome.

Having gone to the trouble of producing a new trainer with good spin capability I think it is a pity the manufacturers did not go a step further and have it cleared for aerobatics. Nevertheless, the Tomahawk handles better than many American light singles, visibility from the cabin is above average, noise levels, if not the quietest are better than some, the cabin is of quite generous size and overall I rate the Tomahawk a good trainer.

at 75 per cent power and 468 nm using 65 per cent. This would allow two adults to be carried without any baggage.

Verdict

During the 1940s and '50s even the big general aviation manufacturers built nothing larger than four-seat tourers and overheads were kept to a minimum. Then faster planes with disappearing legs came along, soon to be followed by light twins, turboprops and even business jets. And with these more complex products came advanced technology, increased overheads and a new problem – the near impossibility of producing simple, two-seat trainers in a factory staffed and equipped to build turboprops or advanced piston twins.

Some have claimed that the only reason companies like Piper or Cessna bother with building cheap trainers on which little or no profit can be made is that people who learn to fly on a particular brand of light plane often buy a larger model of the same make when they become private or corporate owners. Be that as it may, by the time Piper were set fair to launch the Tomahawk, Cessna had built and delivered a staggering 24,000 of their model 150 – so I leave it to the reader to judge whether or not building two-seat trainers is as unprofitable as some people claim.

The Tomahawk is a graceful little aircraft but I thought the standard of finish was not as good as other Piper models. In many areas the metal is paper thin, so that the joints tend to ripple where the top fuselage skin wraps

Facts and figures

Dimensions

Wing span	34 ft
Wing area	125 sq ft
Length	23 ft, 2 in
Height	8 ft, 7 in

Weights & loadings

Max take-off	1,670 lb (758 kg)
Equipped empty	1,156 lb (524 kg)
Useful load	514 lb (233 kg)
Max baggage	100 lb (45 kg)
Max fuel	25 Imp/30 US gal (114 litres)
Wing loading	13.36 lb/sq ft
Power loading	14.9 lb/hp

Performance

Max speed at sea level	109 kt
75% power cruise	108 kt
65% power cruise	100 kt
Range at 75% power	452 nm
Range at 65% power	468 nm
Rate of climb at sea level	718 ft/min
Take-off distance over 50 ft	1,460 ft
Landing distance over 50 ft	1,465 ft
Stalling speed IAS	49 kt
Service ceiling	12,000 ft

Engine

Lycoming 0-235-L2C producing 112 hp at 2,600 rpm, driving a 72-in (183-cm) diameter Sensenich propeller.

TBO

2,000 hours.

Cessna 152

The little survivor from Wichita

Background

It all started in 1911 when an American farmer-mechanic named Clyde Vernon Cessna saw his first plane. Four months later, and after thirteen attempts, he flew in an aircraft built and designed by himself. From those far-away days and through the efforts of Clyde Cessna and, in particular, his nephew Dwane Wallace, the Cessna Aircraft Co has grown to become the world's biggest manufacturer of light and not-so-light aircraft. The range starts with two-seat trainers and covers most shapes and sizes up to business jets.

In 1946 Cessna devised the model 120 and its slightly plusher but almost identical brother, the 140, both of them two-seat trainers with stressed-skin fuselages. Most of them had fabric-covered wings but they finished the production line with 521 of the metal-winged 140A. The last one left the factory in February, 1951. By then some 7,500 of Cessna's little taildraggers had been produced – a success story by any standards – and it is therefore a little surprising that seven years were to pass before the company entered the two-seat trainer market again.

Over those seven years Cessna developed and introduced their model 172, a four-seat tourer with a nosewheel undercarriage; when the new trainer arrived, it was more or less a scaled-down version of the 172. They called it the model 150.

The original Cessna 150 had an upright fin and rudder, mechanically operated flaps and a deep rear fuselage which precluded rearward visibility. Over the years a number of improvements, some of them important, were gradually to transform the little 150 into a useful and civilized trainer. In 1964 the rear fuselage was reduced in depth to allow for a large rear window giving near 360-degree visibility, a swept fin and rudder followed in 1966, improved wing-tips appeared in 1970, and the now-familiar tubular spring steel mainwheel legs came a year later. Inside the aircraft, revisions to the trim managed to reclaim a little width (the 150 is tight in this respect), styling improved, the instrument layout was made to conform with the standard basic 'T', an ignition/starter key replaced the old toggle that had to be pulled, and the flaps became electrically operated. They also offered an Aerobat version with a slightly beefed-up airframe although, frankly, it was not a good aircraft for aerobatics. The ailerons were (and still are) too ineffective and a control wheel is anything but ideal while flying Cuban eights and the like.

By the time production ceased in 1977, the best part of 24,000 Cessna 150s had been built, 1,758 of them at Reims Aviation, a French concern manufacturing some of the Cessna designs under licence. Whether you like the aircraft or not – and the 150 has its critics – 24,000 trainers is a lot of rivets. Clearly one would need to take courage before replacing such a successful product with something radically different. In fact Cessna, quite rightly in my view, decided not to play the hero. Instead, they introduced the model 152 which is really a 150 airframe with very minor changes, the main difference being the replacement of the 100-hp Continental engine with the slightly more powerful Lycoming unit as used in the rival Piper Tomahawk (see previous test).

Engineering and design features

The Lycoming 0-235-L2C resides behind a two-piece cowling which carries the cooling baffles. Removal of the cowlings reveals all and allows for easy servicing. Other changes from the previous 150 model are a 28-volt electric system in place of the original 14-volt one, a 69-in (175-cm) diameter propeller with thinner gull-wing profile blades (claimed to reduce noises), flush window catches and an improved 'follow-up type flap operating switch – you set whatever depression is required, then the flaps move to that position and stop.

The wing centre section is of constant chord but the outer panels are tapered. They carry rather crude ailerons over their entire length. Inboard of the ailerons are aerodynamic devices which Cessna describe in their more unreserved moments as 'Para-Lift' flaps (there is no such animal), while in their excellent flight manual they become 'single-slotted' flaps. In fact the flaps move back on tracking by about 8 in (20 cm) and increase the wing area as they begin to depress. Any flap which does that has surely got to be of the Fowler type. Cessna are great ones for corny trade names dreamed up by ad men who should be put on the pills. Thus they introduce a rear cabin window and call it 'Omni-Vision' while their elegantly simple landing gear is branded a 'Land-O-Matic'. Pay no attention to this childishness. More important is the flaps themselves – in my view the best in the light-plane business.

A fuel tank in each wing gravity feeds to the engine bay and then to an engine-driven pump, there being no need for an emergency electric booster. Standard capacity (usable) is 20.5 Imp/24.5 US gal (93 litres), but an optional long-range system, 31.5 Imp/37.5 US gal (142 litres), is available. The tanks are filled through two points on top of the wing, making it necessary to climb up for the purpose. A small opening in the left-wing leading edge samples air at high angles of attack and near the stall a reed-type warning device lets out a note of protest which changes pitch as the stage is reached where the wing runs

Cessna 152. Simple maintenance and docile handling are attractive features of this very successful light trainer.

The instrument panel is simple, convenient and uncluttered, if a little dated in concept.

Visibility in the air is good except during turns when, like all high-wing designs, the inner wing lowers and obscures the view.

out of magic – simple but effective. The wings are braced to the fuselage by a pair of single, wide-chord struts.

A long dorsal fin runs from behind the rear window to join with the swept fin. A conventional fixed tailplane carries separate elevators which have a large trim tab on the right-hand side. Wheel fairings enclose the brakes but the nosewheel torque links are left out in the breeze where they can only encourage drag.

The 152 is an attractive little machine on the ground with graceful, flowing lines and a lean look to the fuselage.

Cabin and flight deck

Two car-type doors of fairly generous size are provided, one on each side of the cabin. There is also a step on each mainwheel strut to assist in entering the cabin. Leaving the aircraft is less easy because you have to go 'foot-fishing' to find the little step. The cabin is bright and cheerful but my main criticism has always been its narrow width, which at 36 in (91 cm) is 7 in (18 cm) less than a Tomahawk (see previous test). There is a large area behind the two seats where up to 120 lb (54 kg) of baggage can be stowed or, as an extra, a small bench seat may be installed for two children of that weight when holding hands.

A standard basic 'T' flight panel is positioned in front of the left-hand (student's) seat, an rpm indicator, ammeter, and, if required, second altimeter, and flight time recorder take up the right-hand panel, which also carries a small map compartment. The centre of the panel is available for avionics and their readouts (VOR/ILS, ADF, etc).

A panel strip below the main area runs the full width of the cabin. It accommodates the parking brake knob, primer, master switch, two fuel gauges, oil temperature and oil pressure gauges, starter and ignition key switch, a row of rockers for the electric services, carburettor heat knob, a plunger-type throttle, a similar mixture control, the follow-up flap switch and its position indicator, a row of fuses, the cabin heat and fresh-air controls. A small extension to this panel drops down from the centre to provide a home for the elevator trim wheel and its position indicator. There is also a hand microphone for the radio.

In the glareshield is a small audio panel used for controlling which radio is to be heard through the cabin speaker or the headset, a rear-facing mirror (are we being followed!) and, if fitted, the marker beacon lights for the ILS. The right-hand side window opens out and up to provide clear vision. Toe brakes are standard.

Other than the somewhat narrow cabin the interior is well planned and the 152 has a happy, friendly feel to it.

In the air

Taxiing is easy but the nosewheel steering is via bungee chords which link the front strut to the rudder pedals. An anti-shimmy damper is fitted to prevent speed wobble but to this day the 150/152 exhibits traces of this. Up to 8½ degrees of steering, left and right, may be accomplished through the rudder pedals, use of brake increasing the turning angle to a maximum of 30 degrees.

Take-off is totally unremarkable. The 152 runs straight and true without fuss, and I timed lift-off at 13 seconds. Initial rate of climb at 80 lb (36 kg) below maximum weight was a useful 740 ft/min – bettering slightly the previous model 150. The manufacturers claim a 75 per cent power cruise of 107 kt at 8,000 ft at maximum weight and a 65 per cent cruise of 100 kt. The plane which I tested bettered those figures by a couple of knots.

At high angles of attack there is a lot of position error on the ASI so indications tend towards the ridiculous during stalling. Flaps-up we hit the 'g' break at 35 kt and maximum-flap (restricted on the 152 to 30 degrees, as opposed to 40 degrees on the earlier model) showed an uncalibrated position on the dial which I estimated to be 30 kt. Both figures are about 10 kt below reality. I liked the flap operation but thought the switch was flimsy and ready to break at the slightest excuse.

Handling has never been a strong point with the lower-priced Cessna models, partly because of the poorly designed ailerons which provide a roll that even their best friends would not describe as 'crisp'. The 152 is slightly better than the 150 in this respect but lateral control is below average.

There is very little lateral stability in spite of the high wing but only 1½ cycles were required to regain a trimmed 100 kt after a 20-kt disturbance.

The spin is excellent, unhurried and unlikely to upset the student. Entry is smooth applying the rudder at 45 kt, and recovery using the standard method is very satisfactory.

Initial approach may comfortably be flown at 50 kt, aiming for 55 kt over the threshold. There is plenty of elevator power to ensure a mainwheel first arrival – important on this aircraft because constant landings on the nosewheel will eventually lead to failure of the strut. This is true of most aircraft but a particular point to watch on the 152.

Capabilities

There is little to choose between the Cessna 152 and the Piper Tomahawk in terms of performance. Maximum weight is the same in each case: 1, 670 lb (758 kg). An equipped 152 has an empty weight of 1,118 lb (507 kg), leaving a useful load of 552 lb (250 kg), which is 38 lb (17 kg) more than the Tomahawk. If you fill the tanks of the standard version 405 lb (184 kg) is left for payload – say two 170-lb (77-kg) occupants and 65 lb (30 kg) of baggage. So loaded it would fly a with-reserve 350 nm at 75 per cent power. The long-range tank version will fly for 580 nm with a payload of 336 lb (152 kg), which is adequate for two good-sized adults.

Verdict

Although the Piper Tomahawk is, in my view, more pleasant to fly and has a roomier cabin, generally the Cessna 152 gives the impression of being a slightly better piece of engineering.

The Cessna 152 is a reliable, docile little aircraft, a little snug in the cabin and, like most high-wing designs, very blind while turning. I admire the standard version for its simplicity, friendliness, reliability and outstanding success. It withstands life at the flying schools pretty well. But the acrobatic version leaves me cold. Why on earth fit ashtrays in an aircraft capable of slow rolls? (How do you like your cigarette ash and dog-ends – in your hair or down your neck?)

Facts and figures

Dimensions

Wing span	33 ft, 2 in
Wing area	159.5 sq ft
Length	24 ft, 1 in
Height	8 ft, 6 in

Weights & loadings

Max take-off	1,670 lb (758 kg)
Equipped empty	1,118 lb (507 kg)
Useful load	552 lb (250 kg)
Max baggage	120 lb (54 kg)
Max fuel: Standard	20.5 Imp/24.5 US gal (93 litres)
Optional	31.5 Imp/37.5 US gal (142 litres)
Wing loading	10.5 lb/sqft
Power loading	15.2 lb/hp

Performance

Max speed at sea level	110 kt
75% power cruise	107 kt
65% power cruise	100 kt
Range at 75% power: Standard fuel	350 nm
Long-range fuel	580 nm
Rate of climb at sea level	715 ft/min
Take-off distance over 50 ft	1,340 ft
Landing distance over 50 ft	1,200 ft
Stalling speed (IAS)	30 kt
Service ceiling	14,700 ft

Engine
Lycoming 0-235-L2C producing 110 hp at 2,550 rpm, driving a 69-in (175-cm) diameter propeller.

TBO
2,000 hours.

Aerospatiale Rallye Galopin

A tin parachute of a trainer

Background

Governments that are air-minded to the point where they actually *encourage* aviation are very much in a minority. One such rare species was a product of post-war France. In an effort to start a French light-aircraft industry and give the flying clubs a shot in the arm, during the early 1950s the government organized a national design competition for an economic trainer. Much emphasis was placed on safety; some would say too much. The winning design, announced in 1958, was so protective that it made the average light plane seem like a hot rod.

Winner of the competition was the old Morane company. A series of mergers and takeovers resulted in Morane's light plane being built by the Socata division of Aerospatiale, the giant French nationalized aerospace concern that partnered British Aerospace in Concorde, takes a leading role in Airbus Industries and builds, among many other things, some fine helicopters. The light aircraft are manufactured at Tarbes near Lourdes, within sight of the formidable Pyrenees mountains.

The winning design was called the Rallye. At first it was offered with a choice of 90-, 100- or 145-hp Continental engines, but over the years a lot has happened to the breed – engines of up to 235 hp have been fitted. But while the Rallye's remarkable slow-flying capabilities have remained more or less intact, there are so many flap and other brackets hanging off the aircraft that as far as cruising speed is concerned adding power is like throwing folding money down the drain.

Now there is a new range of touring aircraft coming out of the Tarbes factory and this will be described in the next chapter. However, a reduced range of Morane's old Rallye aircraft continues in production and the subject of this report, the Rallye Galopin, is intended as a cheap, rugged, two-seat trainer with a most forgiving nature. Whatever else one might say about the Galopin few would accuse it of being a good-looker. In appearance the Rallye Galopin can best be described as the Deux Chevaux of the air.

Engineering and design features

With safety foremost in the terms of reference for the original design competition that inspired the Rallye range, a number of interesting mechanical and aerodynamic features are built into the aircraft. Structurally the airframe is typical of its period in consisting of many small light-alloy parts riveted to form larger assemblies – a labour-intensive way of doing things. It is stressed so that in a heavy landing the structure collapses from the wheels up, progressively dissipating shock, and cushioning the main spar and fuse-

lage, so far as it possible, against distortion or other expensive damage.

The wing has a small area – only 132 sq ft as compared with the 159.5 sq ft of the Cessna 152. However, it is provided with large-chord autoslats which cover the full span. These are aided and abetted by large-area Fowler flaps which are electrically operated. Pilots unfortunate enough to be following a Rallye flier intent on making a slow approach invariably give up sweating on the edge of the stall and go round again for another try. The Fowler flaps move back on large external tracks – an untidy solution to the problem compared with the Cessna arrangement where the track disappears inside the flap when it is in the raised position. The wide-chord ailerons are also attached by large, external brackets and all this ironmongery gives the Galopin a somewhat untidy appearance.

Autoslats and Fowler flaps endow the Rallye range with remarkable slow-flying capabilities.

Aerospatiale Rallye Galopin. A tough, very safe and reliable trainer with aerobatic potential.

Each wing contains a fuel tank and total capacity is 23 Imp/27.5 US gal (104 litres).

A very lage fin supports an even larger horn-balanced rudder. Mounted on the fin some 6 in (15 cm) above the top of the fuselage is a tailplane of quite massive proportions for a small aircraft. Large elevators carry mass balance weights within bulbous horn balance areas that extend forward of the hinge line. A trim tab protrudes behind the trailing edge of the right elevator. One might question the need for such large tail surfaces on a small trainer but it has been designed for safe, controllable flight at low airspeeds and there is no substitute for area when the knots are few.

The fuselage is rotund and highly tapered, and looks more portly than it is because the wing is attached part way up the sides, leaving a small belly underneath. The engine is the same type of Lycoming as that used in the Tomahawk and the Cessna 152 (see previous airtests). It resides under nicely designed cowlings which may easily be removed by undoing eight quick-release screws.

The entire flying hardware stands on a nosewheel undercarriage of the trailing-link type. The nosewheel is fully castoring, there being no linkage to the rudder pedals, so turns on the ground require the use of differential brake.

Cabin and flight deck

A fixed step is built onto each side of the fuselage to allow the inmates an easy climb into the wing walkways. A handle on top of the large canopy must first be rotated, then the entire transparent area excluding the windscreen slides back to reveal most of the cabin area. Early models in the Rallye range had a rather untidy canopy made up of smaller panels and metal frames. I have known them to jump the sliding track and cause one hell of a noise of hissing air. Current aircraft have a much improved canopy with large, one-piece side panels and a wraparound roof.

The Galopin has two separate seats for student and instructor. There is also a small bench at the rear for two children or one adult and the front seats tip forward to allow them in. Maximum weight at the back is a generous 154 lb (70 kg). Access to the cabin is easier than in many other light planes.

Aerospatiale have made a number of improvements to the Galopin cabin compared with the earlier Rallye models. You needed a degree in engineering to work the original manual flap lever. That has now been replaced by an electric control. In the original version the parking brake took the form of two little levers dangling from the rudder pedals, which threatened to give you double curvature of the spine, even if the straps were slackened first. Now there is a conventionally located knob. Likewise it was necessary to slacken the harness before the fuel selector could be reached (it was on the floor in the old models). That has given way to a sensibly placed fuel knob.

The instrument panel is deep enough for three levels of dials and adequate stowage in the radio area. The rpm indicator is on the right, and in the centre of the panel

below the radio stack is a row of six vertical readout instruments showing the usual engine temperatures and pressures as well as electric charge.

Two plunger-type throttles are provided, one in the centre and another on the left, so a student can be trained to use the throttle with either hand. The electric switches, circuit breakers, ignition/starter key switch and cabin air controls are on a strip which runs below the main panel and extends the full width of the cabin. A wide slot is cut into the right-hand panel to provide a large shelf for flight manuals and so on, while low down on the cabin walls, within easy reach of the pilot's hands, are two large map

The profusion of brackets and tracking for the flaps contribute to high drag and naturally damage the cruise performance.

pockets – a facility that is too often forgotten in modern light aircraft.

At seat level the Galopin measures 39 in (99 cm) across and there is 42 in (107 cm) of shoulder room measured from one side of the canopy to the other. This is not the smartest cabin I have seen on a two-seat trainer, but the Galopin is designed very much down to a price. However, it does offer such goodies as autoslats, Fowler flaps and, believe it or not, pretty good aerobatic capabilities. I say 'believe it or not' because, to be honest, it does not look the part.

In the air
You start up in the usual way, using the electric pump to raise the fuel. Immediately you are aware of the excellent visibility in all directions except for directly behind the aircraft. Although there is no linkage between nosewheel and rudder pedals you soon become used to the techniques of giving a touch of brake in the required direction. On a calm day quite positive changes in direction can be made using slipstream in conjunction with the big rudder.

Some very short take-off distances are possible using the appropriate techniques, but a standard departure will have you at the 50-ft screen in only 1,370 ft on a windless day.

Although the nosewheel is free castoring, the aircraft runs good and straight when you open up power. At about 55 kt you ease back on the stick (control wheels are not fitted in the Rallye Galopin), then as the angle of attack increases, centre of pressure on the slats moves forward and pulls them several inches away from the mainplane with a decisive clonk. Only when the nose is lowered to level out do the slats slam shut.

Initial rate of climb is a sedate 630 ft/min but at 75 per cent power the 104-kt TAS is not a lot slower than the Tomahawk or the Cessna 152 although it gives away 5 kt on the American aircraft's 100-kt 65 per cent power cruise. Noise levels are lower than average and quite acceptable – surprising considering all that clear plastic around the cabin – and visibility during turns and most phases of flight is excellent.

Stability in roll and yaw is good but the Galopin required two cycles to regain a trimmed 95-kt IAS after I had fed in a 15-kt disturbance. In my list of requirements for a good trainer at the beginning of this chapter one desirable characteristic was that it should not be too easy to fly accurately – a little instability is not necessarily a bad thing in this type of aircraft. On the same subject, correct use of rudder is needed to maintain balanced flight at anything other than cruising power and speed. The ailerons are reasonably good but spoiled a little by a pronounced break-out force. The elevators are rather heavy and for my liking the trim wheel (mounted on a small console dropping down from the centre of the instrument panel) is a little too high geared.

Stalling is a new and entertaining experience. Clean, and with the 8-degree take-off setting, the stall occurs at about 40 kt to the accompaniment of those slats thrusting forward from the wing's leading edge like clutching hands. With maximum-flap (30 degrees) there is no clearly defined 'g' break and the tin parachute sinks, wings level, at 36 kt with a 600 ft/min descent indicated on the VSI. I suspect that if you flew it into the ground like that the undercarriage would just about absorb 10 ft/sec.

They claim you can spin if entry is made at 54 kt but I couldn't achieve a proper one, not even with pro-spin aileron, and I suspect a sustained spin is not possible because there is no locking device on the full-span autoslats. As an aerobatic trainer the Galopin is adequate for basic manoeuvres – better, in fact, than the Aerobat version of the Cessna 152. Barrel rolls, loops and stall turns are all entered at 115 kt and it will fly a respectable slow roll.

A comfortable target threshold speed is 65 kt for training purposes but when a short-field arrival is necessary some quite remarkable performances are possible using very low final approach speeds. In the right hands 40–45 over the hedge is comfortable. Crosswind limits are a generous 20 kt and the aircraft is very easy to land.

Capabilities

Maximum take-off weight is 1,695 lb (769 kg) and an equipped Galopin would have an empty weight of around 1,195 lb (542 kg), leaving a 500-lb (227-kg) useful load. This is just about enough for full tanks and two adults without any baggage, a payload it will fly for a with-reserve range of 400 nm. Any load placed on the back seat must be paid for in reduced fuel. With two 170-lb (77-kg) adults up front and the maximum 154 lb (70 kg) in the back you are down to 1 gal in the tank! On the other hand, two adults of average weight 154 lb (70 kg) plus a 126-lb (57-kg) lady in the back would allow enough fuel for a safe 130 nm.

Verdict

The Galopin may be something of an ugly duckling but it is a tough, honest trainer, cheaper than most, and better than any other when the local airfield is of postage-stamp proportions.

As a cheap, reliable two-seat trainer with aerobatic capabilities the Galopin must be rated a success: approaching 3,500 of all models in the range had been built up to the end of 1982.

Facts and figures

Dimensions

Wing span	31 ft, 11 in
Wing area	132.2 sq ft
Length	23 ft, 9 in
Height	9 ft, 2 in

Weights & loadings

Max take-off	1,695 lb
Equipped empty	1,195 lb
Useful load	500 lb
Max load in rear	154 lb
Max fuel	23 Imp/27.5 US gal (104 litres)
Wing loading	12.82 lb/sq ft
Power loading	15.4 lb/hp

Performance

Max speed at sea level	110 kt
75% power cruise	104 kt
65% power cruise	95 kt
Range at 75% power	400 nm
Rate of climb at sea level	630 ft/min
Take-off distance over 50 ft	1,370 ft
Landing distance over 50 ft	830 ft
Service ceiling	10,500 ft

Engine

Lycoming 0-235-L2A producing 110 hp at 2,600 rpm, driving a two-blade, metal propeller.

TBO

2,000 hours.

Robin DR400/120 Dauphin

The wooden wonder from Dijon

Background

I was hard pressed to decide whether or not to include this little aircraft among the two-seat trainers because, although it uses the same Lycoming engine as a Tomahawk or a Cessna 152 (see the first two test reports in this chapter) it manages to fly four adults and cruise faster than all three aircraft previously described. But it can be used, and indeed is used, as a trainer. So on with the story.

It all started in 1948 when two Frenchmen without engineering training decided to design and build their own light plane. Parts were drawn full-scale on some pre-war aircraft-quality plywood found at the back of a hangar. These were the only plans, so as the parts were cut and glued together all drawings disappeared.

Either by accident or intent the partners, Messrs Joly and Delemontez, had settled on a wing plan form that provides the best possible lift envelope. The little single-

seater had a wide flat centre section, with all dihedral confined to the highly tapered and steeply up-swept outer panels, giving it a marked, cranked-wing appearance. It was called the D9 Bébé, and so successful was the first flight that soon a number of other, larger designs followed, all of them based on the same cranked wing.

By 1950 the son-in-law/father-in-law partnership had joined names to form the Jodel title that graced so many outstanding light planes flying around Europe – so many because, instead of going into production, Edouard Joly and Jean Delemontez preferred to license the Jodel concept and allow others to build developments based on their original design. Among several small manufacturers building Jodels in France none has taken the Jodel concept to a higher degree of perfection than Pierre Robin, one-time flying instructor at the little airfield of Darois: a rough, grass surface situated almost 2,000 ft above sea level in

Robin DR400/120 Dauphin. Note the clean engine installation, the extended air intake and the forward-sliding canopy.

beautiful countryside near Dijon in central France.

Avions Pierre Robin is a small modern factory with metal aircraft under construction in the main part, and a quiet, tranquil shop building the Jodel-inspired models out of wood and Dacron nearby. A steady demand exists for these wood-and-Dacron masterpieces, mainly from discriminating pilots who know that modern wooden construction can outlast metal. There are no corrosion or fatigue problems and there is less engine noise and vibration transmitted by the airframe. Certainly they stand up to the rough treatment meted out by the flying schools.

Engineering and design features

The aircraft in this test report is called the Dauphin, which is the lowest powered of Pierre Robin's DR400 series. It is the practice at the Dijon factory to quote the engine power after the model number and sometimes (but not always) give the type a name. So the Dauphin is the DR400/120. There is also a DR400/160 and a DR400/180 (160 hp and 180 hp respectively).

To deal with the engine first, it is basically the same unit as the Lycoming 0-235 as fitted in the Tomahawk, Cessna 152 and Rallye Galopin. But by using Bendix magnetos in place of the original equipment, it has been possible to increase maximum rpm from 2,600 to 2,800 and add about 8–10 hp into the bargain. The result? An astonishing little aircraft with enough useful load for four adults.

The wings are based on a single adequately proportioned box spar, ply covering being used for the front portion of the mainplane. Behind the spar, Dacron covers are used (they have an unlimited life, by the way). A favourite trick of one Robin salesman is to throw a bunch of keys onto the fabric and watch them bounce off without leaving a mark. Imagine doing that to a metal wing!

Simple, rather basic-looking ailerons are fitted to the tapered, outer wing panels and the flat, parallel-chord centre section carries plain flaps, which are the least attractive feature of this otherwise outstanding light plane.

The graceful fuselage is entirely ply covered and the top longerons are low set where they form convenient arm rests in the cabin. Massive side windows extend up from that level and an unusual feature is the large, one-piece Robin canopy, which includes the windscreen. Like the side windows it extends from elbow level, up and over the roof, and the entire clear plastic area slides forward to reveal much of the cabin for easy entry and exit. The advantage of this arrangement is that the canopy closes in smooth conformity with the fuselage. There are no gaps or joints to cause drag.

Standard fuel capacity is 24.4 Imp/29.5 US gal (112 litres) carried in a single fuel tank located under the baggage area at the back of the cabin but an optional 36-Imp/43-US-gal (163-litre) system is available. The engine resides within well-designed fibreglass cowlings and the carburettor air intake is built forward to keep it away from stagnant airflow, thus ensuring good induction. At the other end of the fuselage a nicely proportioned fin blends elegantly with the lines of the aircraft. The stabilator is provided with an antibalance tab which is adjusted in datum by the elevator trim wheel.

Although the undercarriage is fixed, Pierre Robin turns out the best wheel fairings in the business – even the torque links on the nose strut are carefully enclosed. It is attention to such detail that gains him a few knots here and a few feet per minute there.

In terms of finish, Robin aircraft are beautifully turned out and viewed on the ground the Dauphin is a pleasure to look at.

Cabin and flight deck

A fixed step is provided on each side of the fuselage and the aircraft can be entered from both sides using the two wing walkways. A single lockable handle on the roof is turned through 90 degrees to allow opening of the canopy, which glides smoothly forward on nylon runners. The front seats tip forward, allowing easy access to the rear bench.

The cabin is fractionally wider than a Piper Warrior, with a surprising amount of leg room at the back. Excellent support is provided by the two pilots' seats which adjust for leg reach, raising slightly as they are moved forward.

The instrument panel has ample room for a full basic 'T', airways radio, engine readouts, circuit breakers, electric switches (on a full-width strip below the main panel), and an annunciator (warning/alerting lights) built into the glareshield. Yet the panel is set low enough to allow unobstructed view over the nose – a masterpiece of good design that seems to be beyond the wit of the big manufacturers of light aircraft. In fact, visibility from this aircraft in most directions is in a class of its own and perhaps only equalled in some helicopters.

Two throttles are provided, one in the centre and another on the left to allow training for left- or right-hand throttle management. It is useful to be able to adjust throttle friction according to circumstances. However, these throttles are pre-tensioned with no friction adjustment, a feature that frankly I do not care for. A small console between the seats carries the trim position indicator, a large brake handle which is also used for parking, a fuel selector, the pleasant-to-use trim wheel and a flap lever. Control sticks bend away under the instrument panel to cause a minimum of obstruction in the cabin.

Good map pockets are provided and there are four adjustable fresh-air vents. Two more document pockets are located in the front seat backs. Generally the cabin is smart, comfortable and nicely finished.

In the air

For airtests I like, whenever possible, to fly the aircraft at or near its maximum weight. On this occasion there was a man on my right who admitted to 215 lb (98 kg), while a young chap who stood at least 6 ft, 4 in (193 cm) sat in the back. He said there was plenty of room and in total I estimate the three of us weighed 550 lb (250 kg). Although

Robin instrument panels are well planned. The two comfortable seats give excellent support and adjust for leg reach.

we had full tanks there was 50 lb (23 kg) available for baggage.

Excellent disc brakes are fitted and these are applied with the large parking brake handle. The nosewheel steering may be augmented by applying full rudder when the brake on that side comes on to assist the turn. On the ground a Dauphin is easier to drive than a family car.

From full throttle to lift-off was fourteen tranquil seconds. There was no tendency to swing, hop or porpoise. Initial rate of climb was 700 ft/min and this settled at the book value of 600 ft/min. When used as a trainer with only two people on board the aircraft goes up quite rapidly, 750–800 ft/min being average.

You can order the Robin Dauphin with a coarse-pitch propeller. It works wonders for the cruise without damaging climb performance too seriously, and at 7,000 ft the 75 per cent power setting gives a cruising speed of 116 kt. At maximum weight you can achieve a continuous 120 kt without trouble. Up to 2,600 rpm the aircraft is quieter than average – about acceptable – but noise levels naturally rise at higher revs.

Without flap the aircraft stalls, wings level at 50 kt and full-flap reduces that by 5 kt. A slight tendency for the left wing to drop was corrected automatically by the aircraft without my taking any action.

The ailerons are effective but a little heavy and generally

the Dauphin is a delight to fly, with the most outstanding visibility for all occupants. Stability is adequate in all three axes.

For training purposes the DR400/120 is cleared for spinning with two people on board but due to the aircraft's characteristics it is only possible to demonstrate the entry. After about half a turn the airspeed begins to increase as the spin converts to a spiral dive. Although this is a pity, the plus marks are earned by the Dauphin's higher cruising speeds than the other trainers' and its ability to take four adults.

Initial approach is flown at 70 kt aiming for 60 kt over the runway numbers. They do not come any easier to land than the Dauphin.

Capabilities

Bearing in mind that the Robin DR400/120 contrives to fly faster than the Tomahawk or Cessna 152 on virtually the same engine, and provides more than 200 lb (91 kg) of extra useful load in the bargain you are entitled to ask 'How is it done?' In the first place Pierre Robin takes more care over the drag-producing details than the big manufacturers. Second, the Jodel wing carries heavy washout at the tips so that in cruising flight most of the lift comes from the centre section with very little drag being generated by the outer panels.

The DR400 series is based on a particularly efficient airframe and an outstanding wing.

Assuming the larger-tank version, you could take on maximum fuel and carry three large adults over a with-reserve range of 665 nm at 75 per cent power. Alternatively the DR400/120 can carry four 160-lb (73-kg) adults and 50 lb (23 kg) of baggage over a safe 200 nm, in motoring terms returning 24 mpg (Imp gal) while cruising at 133 mph. Beat that in your best Ferrari!

Verdict
For its power the Robin DR400/120 is truly remarkable, putting to shame some tourers with the advantage of another 40 hp. You can see DR400 aircraft of various engine powers at many flying schools throughout Europe, and elsewhere. Some are twelve or more years old yet they retain their smart appearance. The Robin is a tough little bird — not just pretty and fast.

Facts and figures

Dimensions

Wing span	28 ft, 8 in
Wing area	146.5 sq ft
Length	22 ft, 10 in
Height	7 ft, 4 in

Weights & loadings

Max take-off	1,984 lb (900 kg)
Equipped empty	1,213 lb (550 kg)
Useful load	771 lb (350 kg)
Max baggage	88 lb (40 kg)
Max fuel: Standard	24.4 Imp/29 US gal (110 litres)
Optional	36 Imp/43 US gal (163 litres)
Wing loading	13.5 lb/sq ft
Power loading	16.81 lb/hp

Performance

Max speed at sea level	128 kt
75% power cruise	116 kt
65% power cruise	108 kt
Range at 75% power	450 nm
Range with long-range tanks	665 nm
Rate of climb at sea level	600 ft/min
Take-off distance over 50 ft	1,750 ft
Landing distance over 50 ft	1,500 ft

Engine
Lycoming 0-235-L2A developing 120 hp at 2,800 rpm, driving a two-blade Sensenich propeller.

TBO
2,000 hours.

Robin R2160

A shot in the arm for flying training

Background

Pierre Robin made his name building the type of aircraft described in the previous airtest. And with the advent of modern glues, indefinite-life Dacron and tougher paints his beautiful wooden wonders became better and better. However, the mania for metal spread on a worldwide basis despite certain real disadvantages where light planes are concerned, prejudice against wood set in, and although the DR400 series continues in production (two of the range are described in this book) the Robin concern felt compelled to enter the aluminium age.

First metal design, which appeared in 1970, was the HR100 (H standing for Christophe Heintz, and R for Pierre Robin – the two designers). It was a four-seat tourer with a range of 1,500 nm and a fuel capacity to match:

100 Imp/120 US gal (454 litres) no less. Soon afterwards came the HR200, a delightful little two-seat trainer. Originally it was intended to have it certified for spinning and aerobatics but early indications were that although spin entry was very difficult the subsequent recovery might prove impossible. Some flying instructors managed to prove this in Britain by adopting an unauthorized entry method; a flat spin ensued and although badly injured they lived to fly another day.

The subject of this test report is a development of the HR200, which was introduced in 1971. Robin improved the ailerons, greatly increased the size of the fin and rudder, carried out exhaustive spinning tests (adding a large ventral fin as a result) and called it the R2100. By now

Robin R2160, an all-metal two-seat aerobatic trainer offering a wide cabin of outstanding comfort.

the design was the sole work of Pierre Robin. Two models were offered, the R2112 with a 112-hp Lycoming engine and the R2160 which, as you will have guessed, enjoys the urge of a 160-hp motor.

Avions Pierre Robin have opened a plant in Canada where the R2160 and future models can be assembled. In 1982 the R2160 became Robin's first aircraft type to receive certification in the United States.

Engineering and design features

The R2160 is a compact little plane with a wing span of only 27 ft. Well-designed slotted flaps and Frise ailerons are fitted and, unusual these days, there are no tanks in the wing, all fuel being carried in a 26-Imp/31-US-gal (117-litre) cell located under the luggage shelf behind the two seats. The filler point is on the left side at the back of the canopy. The wing is of parallel chord and has about 6 degrees of dihedral.

A Lycoming 0-320-D engine is hidden behind close-fitting fibreglass cowlings which may quickly be removed to reveal all. Like other aircraft from this manufacturer great care is taken to project the carburettor air intake

well forward of the nose to avoid stagnant airflow.

The cabin area is covered by a large forward-sliding canopy which takes with it the windscreen and leaves behind the rear portion covering the baggage area. Unobstructed visibility is provided through 360 degrees. There is a stabilator (all-flying tailplane) with a pair of anti-balance/trim tabs and a very large rudder hinged to a swept fin. The only feature to spoil an otherwise appealing aircraft is the ugly ventral fin which has had to be added as an insurance against non-recovery from a spin. The plane stands on a nosewheel undercarriage with oleo shock absorbers and probably the best wheel fairings you have ever seen. After take-off the nose strut extends, disengages from the rudder pedal steering and aligns the wheel spat with the aircraft's fore and aft axis. The airframe is stressed to +6g and −3g.

Cabin and flight deck

To assist entry and exit a step is provided behind each wing root. Standing on the left or right wing walkway it is easy to reach the canopy handle (located top centre), twist it through 90 degrees and then slide the windshield and canopy forwards. Inside the 42-in (107-cm) wide cabin (sufficient width for a trainer) is a pair of very comfortable bucket seats which give excellent support and are adjustable for leg reach, rising slightly as they move forward. The seats will accept parachutes if these are required and excellent-quality five-strap harnesses are provided with a proper quick-release box – a pleasant change from the rubbish fitted by the big manufacturers, which either threatens to have your head off or tends to undo in the air.

Behind the seats is a large baggage shelf fitted with bungee chords. Up to 77 lb (35 kg) may be stowed in that area, which is reasonable if not excessive for two people.

A comprehensive annunciator is built into the glareshield and standard instrument lighting is concealed underneath. Individual pillar lights may be specified as an option. There is ample space for radio and a proper flight panel. Engine instruments, with the exception of the rpm indicator over on the right-hand panel, are on a strip running below the main area. On either side of the temperature, pressure, electric and vacuum readouts are the switches for the electrics (thermal overload type that pop out of circuit when under duress) and the fuses which are over to the right. A neat and tidy central console drops down and continues back between the seats. It holds a key-type ignition switch, a separate starter button, the battery master and alternator switches, mixture control (like all French aircraft it is the wrong colour – yellow instead of red), flap switch and the position indicator for elevator trim. Further back and within convenient reach of the two seats is the trim wheel. There is a simple ON/OFF fuel selector with the fuel pump switch above it.

Toe brakes are fitted and these may be locked ON for parking by means of a toggle that pulls out of the base of the instrument panel. This is an aerobatic trainer and sticks are fitted. Less happy are the two plunger-type throttles,

one on the left, the other in the centre, which have no adjustable tension. I think Pierre Robin should replace these with throttle levers.

The standard of design and finish, inside and out, is above average and the cabin makes an attractive classroom for student pilots.

In the air

You prime the engine by pumping the throttle, there being no separate plunger. The disc brakes are very powerful, nosewheel steering is excellent (spring linked to the rudder pedals) and all-round visibility is light years ahead of most other aircraft. A compact plane with a lot of power is a sure recipe for snappy performance. I was not disappointed. From opening the throttle to lift-off was only twelve seconds and almost immediately 1,200 ft/min settled on the VSI, a figure I later confirmed on the stop-watch. We were not far below maximum weight for this flight yet the aircraft was going up 180 ft/min faster than claimed by the manufacturer.

At 2,500 ft, 75 per cent power resulted in a TAS of 129 kt. Fully loaded Robin claim 130 kt at 8,500 ft, and 65 per cent power will give a cruise of 127 kt at 11,000 ft. These speeds are high enough for meaningful navigation exercises. Noise levels at the higher speeds are average; 65 per cent makes the cabin much quieter. Clean stall occurred at 55 kt, wings level, and full-flap (35 degrees) provoked a slight tendency to roll off left at 46 kt.

The ailerons are firm but effective, the large rudder is about right, but the elevators could do with being made a little lighter and the gearing to the anti-balance tab seems to need altering. I would describe the handling as heavy but precise. Stability is good in all planes of rotation and Robin have got the conflicting requirements of stability and control in an ideal trainer just about right.

The R2160 has been spin tested to six turns, which is adequate because in my view there is little point in exceeding two or three turns during training. It only upsets the student and proves nothing. I closed the throttle, held up the nose and at 60 kt applied full left rudder while bringing back the stick. Spin entry was smooth and rotational speed was moderate. After two turns full opposite rudder followed by progressive forward movement on the stick resulted in an almost immediate recovery.

Recommended speeds for aerobatics are:

Barrel roll	130 kt
Slow (aileron) roll	125 kt
Loop	135 kt
Stall turn	130 kt
Roll off/Cuban eight	135–140 kt
Flick roll	70–80 kt

Having run through this sequence nonstop, in that order, the accelerometer fitted to the demonstration aircraft registered only 2½ g. The rudder and ailerons feel satisfactory during a routine like this but the elevators are too heavy for comfort. Having said this, the slow roll (aileron roll in the United States), which can be a little

ABOVE: *The forward-sliding canopy is a feature of all Robin aircraft.*

OPPOSITE: *The Robin R2160 has been given a large rudder and generous fin area to ensure good spin recovery.*

untidy in some of the so-called aerobatic mounts, and even a few of the respected ones, is no problem in the R2160.

Visibility during all phases of flight is unsurpassed and this is particularly important while flying in a busy circuit. Initial approach may comfortably be flown at 70–75 kt. I thought the flaps were a little short on drag but this is typical of the slotted variety and I have known worse. Short finals are comfortable at 60–65 kt and the aircraft feels good on the way in to the runway. Pitch control remains effective with adequate power to keep the nose-wheel off the ground after touchdown. Short landings, using a final approach speed of 55 kt allied to the powerful brakes, can be quite impressive.

Capabilities

Maximum weight is 1,770 lb (803 kg), 100 lb (45 kg) more than a Tomahawk or a Cessna 152 but 50 lb (23 kg) less than the Grob G109 (see the next test report). Equipped empty weight would average 1,217 lb (552 kg), leaving a useful load of 553 lb (251 kg). Maximum fuel weighs in at 187 lb (85 kg), allowing 366 lb (166 kg) for student and instructor, which is more than enough. On a touring basis two people of average weight, 160 lb (73 kg), could take 46 lb (21 kg) of luggage.

Range at 75 per cent power is 365 nm increasing to 430 nm at 65 per cent, in each case allowing for the usual reserves. Two 170-lb (77-kg) occupants and the 77-lb (35-kg) maximum baggage allowance would mean reducing fuel by 7 Imp/8.5 US gal (32 litres) when the two ranges

become 240 nm and 315 nm respectively.

Verdict

The Robin R2160 is the most powerful trainer in this chapter. Although many people in the past have learned to fly on similar-powered Cherokees and the like, the inflated cost of fuel has seen a move towards smaller engines. However, there is a place for the more powerful trainers and in terms of customer appeal the R2160 has a lot going for it. Engineering standards are better than those of some manufacturers and if the aircraft does not handle as nicely as its predecessor, the HR200, it nevertheless makes rings around the mass-produced opposition in this respect. 1,770 lb (803 kg) is very light for a 160-hp trainer and to utilize its good cruising speeds I would have thought it possible to increase fuel capacity and provide another 100 nm range. Having said this, if I were foolish enough to run a flying school the R2160 with its comfortable and very robust airframe would be high on my list for consideration.

Facts and figures

Dimensions

Wing span	27 ft, 4 in
Wing area	140 sq ft
Length	23 ft, 3 in
Height	7 ft

Weights & loadings

Max take-off	1,770 lb (803 kg)
Equipped empty	1,217 lb (552 kg)
Useful load	553 lb (251 kg)
Max baggage	77 lb (35 kg)
Max Fuel	26 Imp/31 US gal (117 litres)
Wing loading	12.64 lb/sq ft
Power loading	11.06 lb/hp

Performance

Max speed at sea level	139 kt
75% power cruise at 8,500 ft	130 kt
65% power cruise at 11,000 ft	127 kt
Range at 75% power	365 nm
Range at 65% power	430 nm
Rate of climb at sea level	1,023 ft/min
Take-off distance over 50 ft	1,345 ft
Landing distance over 50 ft	1,361 ft
Service ceiling	15,000 ft

Engine
Lycoming 0-320-D developing 160 hp at 2,700 rpm, driving a Sensenich, 72-in (183-cm) diameter, fixed-pitch propeller.

TBO
2,000 hours.

Grob G109

The shape of things to come?

Background

I end this section on two-seat trainers with a very different light plane, one that owes little to established methods of construction. It is, perhaps, more closely related to high-performance sailplanes than the usual run of light aircraft. Although the original idea was inspired by the self-launching motor glider concept, so practical are some of these aircraft that basic training for the private pilots' licence is being conducted on such equipment at a number of flying schools.

Motor gliders are being built in a number of European countries. There is the Rumanian Brazov IS-28M2 (built in metal), a Polish-built example and several that come from West Germany. The subject of this report is from Grob Werke GmbH, an outfit run by a Swiss who has set up shop in West Germany. Over the years Grob have earned for themselves a good reputation for building high-quality sailplaines out of glass-reinforced plastic (GRP), which is our old friend glass fibre or fibreglass if you

prefer. So it is not surprising that they should stick to the material in which they are so clearly expert in building their motor glider, the G109.

Engineering and design features

The 54 ft, 6 in (16.61 m) wing is based on a carbon fibre mainspar. There are no flaps but a pair of airbrakes or spoilers lift out of the top surface of the wing. These are essential because this remarkable aircraft has a gliding angle of up to 1 in 30, and when thermals abound the G109 wants to go up rather than down, even with the propeller stopped! On the ground there is very little dihedral but like most very high-aspect-ratio wings (13.5 in this case) there is some flexing when the aircraft is in flight and dihedral increases somewhat under load. The wing has large root fillets which blend gracefully with the fuselage. To prevent damage in the event of a wing touching the ground, bumpers are moulded into the tips.

One of the charms of fibreglass is the relative ease of incorporating such intricate shapes into a design.

The fuselage is fairly wide in the cabin area but tapers sharply behind the seats to become a tube which runs back and supports the tail surfaces. It terminates in a large swept fin on top of which is a fixed tailplane with separate elevators. There is a trim tab on the right-hand side. A large tailwheel hides within a nicely designed fairing. It steers through the rudder pedals until one brake is applied to tighten up the radius of turn, when it unlocks itself to become fully castoring through 360 degrees. Spring steel legs attached to the fuselage carry the mainwheels which are provided with hydraulic disc brakes. A small step on the front of each leg enables you to enter the aircraft over the leading edge of the wing.

The all fibreglass Grob G109 enjoys a surface finish that could not be achieved by a metal airframe.

The wide cabin offers adequate room for radio and instruments. The airbrakes are controlled on either of the two levers at the extreme left and right of the cockpit.

The G109 is powered by a Limbach L2000 EJ engine which develops 80 hp. It is based on the old Volkswagen Beetle motor car engine. There are a number of these conversions on the market and many people are surprised at the success of VW motors when they are used to power light planes. However, they were originally intended as small aero engines before the Second World War, prior to their adoption in Hitler's 'People's Car'.

The top cowling may quickly be removed to reveal the beautifully finished engine, which, with its two carburettors and polished metal, looks like that of an expensive sports car. Thrust is provided by a talented Hoffman propeller which is manually set to FINE for take-off, COARSE for cruising and FEATHER for when conditions are suitable for soaring without power.

The standard of finish on this airframe must be seen to be believed. There is not a wrinkle or a blemish anywhere;

the entire surface – from wing-tip to wing-tip and nose to tail – is like a melamine table mat. With such an ice rink of a surface it is hardly surprising that no wing walkways are provided. Instead, entry and exit are made ahead of the wing, using the undercarriage steps already mentioned.

The cabin area is covered by a very large canopy of high optical quality which hinges forward to reveal the two seats. Without doubt the G109 is a thing of beauty and striking to behold on the ground.

Cabin and flight deck

The G109 is no toy. The cabin is fractionally wider than the Tomahawk's and provides 7 in (18 cm) extra room across the seats compared with the Cessna 152 (see previous test reports). And what splendid seats they are with their slightly reclining position. Each pilot has a large airbrake lever by his side of the cabin. They close with a firm clonk and quite a pull is required to start applying the airbrakes. After they come out of the wing upper surface movement is easy and progressive until the final 2 in (5 cm) of movement when the wheel brakes are partly applied – a very convenient arrangement during the landing roll.

The wide instrument panel has plenty of room for the standard basic 'T' but I was intrigued to see a VSI calibrated in knots up and down instead of feet per minute. However, there is no need to have a degree in mathematics to do a rapid conversion; 1 kt UP means 6,080 ft/hour or, for practical purposes, 100 ft/min. Engine instruments are on the right-hand panel with an area in the centre provided for avionics. A small console extends down from the centre of the panel, which carries the cabin heat control, the choke (remember this engine has been pinched from a car), carb heat, mixture control and a pair of propeller controls. One takes the form of a toggle which is pulled out at 2,200 rpm to go from FINE to COARSE pitch. You revert to FINE by repeating the process at 1,200 rpm. The other propeller control is a large handle. For soaring you stop the engine and pull it to select FEATHER.

The console extends back between the two seats and contains the ON/OFF fuel selector for the 17.6-Imp/21-US-gal (80-litre) tank, which resides under the baggage shelf behind the two seats. There is also a brake lever which applies both discs and may be locked in the parking position. The throttle works in a neat little quadrant built onto the console and a pair of short control sticks sprout from the front of the seat squabs. They fall conveniently to hand when you strap yourself into the machinery. Electric switches and circuit breakers are lined up along the bottom of the right-hand panel.

The big canopy pulls down and shut. It is prevented from falling with a thunderous crash by a pair of damping struts. Everything inside this remarkable little aircraft is excellently designed and finished to the highest quality. Inside and out it is a very refined piece of engineering.

In the air

Taildraggers being taildraggers I wondered what to expect, because there was quite a strong wind blowing on the day of my test and I was mindful of that big fin. In the event there were no problems and ground handling was almost up to the best nosewheel standards. So was the view over the nose due to the G109 having quite a shallow tail-down stance.

Engine checks are a bit of a non-event; there is only one ignition switch so there are no mag checks. But naturally you check the temperatures and pressures and, of course, the electric charge. Then it is vital to ensure that the propeller is in fine pitch before attempting the take-off; you should get 3,000 rpm against the brakes, which, incidentally, are excellent.

While on the ground the aircraft stands almost level; consequently there is very little change in attitude while raising the tail during take-off. Hardly any rudder is required to maintain direction and at around 50 kt the stick is brought back to lift cleanly off the ground. Initial rate of climb is in excess of 500 ft/min – acceptable considering the low power and the large, comfortable cabin.

After reaching the required cruising level the rpm are reduced prior to pulling back the propeller toggle and going into coarse pitch when full throttle may be used to attain 3,000 rpm. At low levels, say 3,000 ft or so, this will give you a respectable 108-kt TAS for a fuel burn of only 3.3 Imp/4 US gal (15 litres) per hour. So for the benefit of car owners who are not yet pilots we are talking of almost 38 mpg while cruising at 125 mph in a comfortable cabin for two. I thought that noise levels were quite low but to make further improvements Grob are now fitting a rubber seal around the canopy frame which may be inflated with a little handpump after the lid has been closed. I was particularly impressed with the high quality of the canopy which seemed to be entirely free of distortion. During turns, the top of the glareshield tends to intrude into the forward vision slightly and I would like to see it lowered on future models. Otherwise visibility in most directions is excellent.

There is little lateral stability and not a lot in pitch so the aircraft must be flown all the time – not a bad thing in a trainer but less desirable in a touring aircraft. Possibly the G109 needs a larger tailplane or its angular relationship to the mainplane requires adjusting. All flying controls, trimmer included, are good although the ailerons, which should be the lightest to handle, are heavier than either the rudder or elevators. In an ideal aircraft, control force inputs for the ailerons, elevators and rudder should be in a proportion of 2:4:6, in that order.

The lowest speed at which I was able to fly the G109 was 48 kt, which is quite high for this class of aircraft, but then it has no flaps. At 60 kt, with the propeller windmilling, rate of descent is about 300 ft/min, increasing to 1,000 ft/min when full airbrake is applied. This is spring loaded so that only half airbrake remains extended should the pilot release the lever.

70 kt works well for the approach, aiming for 60 kt over the threshold. You can leave the throttle alone and control the glidepath on the airbrake but for conventional training a better technique is to set the airbrake half open and adjust the throttle in the normal way. But the aircraft is so clean that part airbrake is essential – otherwise the bird will float on forever. Only a small change of attitude is needed to reach the three-point position and you must guard against landing tailwheel first. After touchdown, application of full airbrake dumps the remaining lift and brings on the wheel brakes to shorten the landing roll without need of grabbing the handbrake lever.

Capabilities

Equipped empty weight at 1,280 lb (581 kg) leaves a useful load of 540 lb (245 kg), maximum take-off weight being 1,820 lb (826 kg), which is 150 lb (68 kg) more than a Tomahawk. Full fuel would account for 127 lb (58 kg) and allow 413 lb (187 kg) for the inmates – enough for two 170-lb (77-kg) adults, the 44-lb (20-kg) maximum baggage allowance and another 29 lb (13 kg) for flight manuals, etc. So the Grob is not short on payload.

In the air the graceful lines of the Grob G109 are a clue to its sailplane ancestry.

Maximum fuel will fly a full cabin for 485 nm at 108 kt or 540 nm at 95 kt. These are with reserve ranges and they compare well with the other light planes in this chapter. However, I think the manufacturers should consider using the spare 29 lb (13 kg) of payload by providing a slightly larger fuel tank. That would translate into another hour at 108 kt.

Verdict

As a motorized glider the Grob G109 is in a different category to the other trainers described in this chapter. With a maximum gliding angle of 1 in 30 it is certainly a very efficient powered flying machine.

Assessed as a trainer one must take into account the 54-ft wing. Such a span on a low-slung aircraft can present problems unless operations are confined to an obstruction-free airfield. At the time of my airtest in 1982 Grob were in the process of obtaining spin clearance, but in all other respects I have no doubt the G109 would fulfil most of the training requirements of a PPL syllabus. In these days of high fuel costs the good news has got to be that the G109 equals the performance of a Cessna 152 on 30 less hp while carrying its two occupants in a cabin offering about 7 in (18 cm) more width.

The Grob G109 is not only very economical but fun to fly as well. Standards of workmanship, quality and finish are as good as I have seen. The airframe is wax polished at the factory for added protection. The Limbach engine has a 1,000-hour life in West Germany but in Britain it is operated under the 'on condition' scheme. Could it be that with fuel prices at their present levels this class of aircraft points the way for future trainers? Well Grob obviously think so. Because at time of writing they are developing a two-seat trainer with a similar airframe based on a wing of more normal span. They call it the Grob 110 – and from all accounts it will set new standards in light trainers.

Facts and figures

Dimensions

Wing span	54 ft, 5 in
Wing area	219.57 sq. ft
Length	25 ft, 10 in
Height	5 ft, 6 in

Weights & loadings

Max take-off	1,820 lb (826 kg)
Equipped empty	1,280 lb (581 kg)
Useful load	540 lb (245 kg)
Max fuel	17.6 Imp/21 US gal (80 litres)
Max luggage	44 lb (20 kg)
Wing loading	8.28 lb/sq ft
Power loading	22.75 lb/hp

Performance

Max speed at sea level	119 kt
75% power cruise	108 kt
Range at 75% power	485 nm
Range at 95 kt	540 nm
Rate of climb at sea level	530 ft/min
Best glide	1 in 30
Minimum sink	1.1 m/sec
Take-off distance over 50 ft	1,410 ft
Landing distance over 50 ft	1,280 ft

Engine
Limbach L2000 EJ developing 80 hp, driving a three-position Hoffman propeller.

TBO
1,000 hours in West Germany. 'On condition' in Britain.

HOW DO THEY RATE?

The following assessments compare aircraft of similar class. For example, four stars (Above average) means when rated against other designs of the same type.

★★★★★ Exceptional
★★★★ Above average
★★★ Average
★★ Below average
★ Unacceptable

Aircraft type	Appearance	Engineering	Comfort	Noise level	Visibility	Handling	Payload/Range	Field Performance	Cruising Speed	Economy	Value for Money
Piper PA-38-112 Tomahawk	★★★★	★★	★★★	★★★	★★★★	★★★	★★★	★★★	★★★	★★★	★★★
Cessna 152	★★★	★★★	★★	★★★	★★	★★	★★★	★★★	★★★	★★★	★★★
Aerospatiale Rallye Galopin	★★	★★★★	★★★	★★★	★★★★	★★★	★★★	★★★★	★★	★★★	★★★★
Robin DR400/120 Dauphin	★★★★	★★★★	★★★	★★★	★★★★★	★★★★	★★★★★	★★★	★★★★	★★★★★	★★★★
Robin R2160	★★★★	★★★★	★★★★	★★★	★★★★★	★★★★	★★★	★★★	★★★★	★★★	★★★★
Grob G109	★★★★	★★★★	★★★	★★★	★★★★	★★★	★★★	★★★	★★★	★★★★	★★★

3. SINGLE-ENGINE FIXED-UNDERCARRIAGE TOURERS

In the preface to this book I complained that very little progress in light plane performance had occurred since the golden years between 1935 and 1939. What would we all think of the motoring industry if the automobiles of today were no faster, quieter, more economical and better to drive than, say, a car of the 1930s? Yet with few exceptions that is the existing state of play with many of the single-engine tourers. The trainers described in the previous chapter are all more comfortable, easier to fly and better able to cope with strong winds than pre-war trainers. Likewise the light twins described in Chapter 5 are, in the main, light years ahead of pre-war equipment. But single-engine tourers!

What is required of a good tourer? Well some years ago I was asked by the magazine *Flight International* to write on this topic and in essence I had this to say about my ideal tourer.

- The cabin must be wide enough for comfort, 42 in (107 cm) at seat level being an absolute minimum.
- All essential controls must fall nicely to hand and work in a natural sense – for example, trimmers should wind forward for nose down, and to the right for right rudder. They should be logically grouped. For example, fuel management must be idiot-proof with the selector pointing to the fuel contents gauge of the tank in use; the flap control and its position indicator should be together, and so on.
- All electric circuit breakers must be accessible in flight without the harassed pilot having to stand on his head to be able to find the one that has thrown out of line.
- A good annunciator panel is essential these days, not a luxury to be ordered as an option. The private owner should not be expected to pay extra for the privilege of knowing if the alternator is on the blink or if the flaps have crept down a little.
- There must be adequate stowage for airways manuals and maps. We had big shelves under instrument panels before the war, at a time when no Aerad or Jeppesen manuals existed. Now that even low-powered tourers have the radio to make full use of such manuals few modern tourers provide a space for parking these bulky publications.
- There is little point in providing a generous baggage

allowance of, say, 200 lb (91 kg) if there is not enough space to stow the cases. A baggage allowance of at least 35 lb (16 kg) per occupant should be provided along with sufficient locker space to take it.
- My ideal single-engine tourer would be in no doubt about its purpose. It would either be a four-seater or a genuine six-passenger plane. So-called '4 plus 2' aircraft (where two seats are provided at the rear of the cabin for the children) are a potential danger, because we all have our own idea of when a child becomes an adult. A pair of sturdy 160-lb (73-kg) children sitting in seats intended for a brace of nippers can play Hamlet with the centre of gravity, not to mention the maximum take-off weight. Pilots are people – and people who only fly intermittently are likely to make the kind of mistake I have in mind.
- There must be adequate ventilation and the heating system should be capable of separate temperature adjustment for the front and rear areas of the cabin. The boys up front are often too hot while in my experience the girls in the back almost invariably complain of feeling cold.
- Noise levels must be low. Little progress has been made in this direction because the manufacturers hardly ever take noise seriously. Yet it is an established fact that noise causes fatigue. Those who fly light singles regularly have to develop good lungs to engage in conversation while in the air. Often the exhaust terminates under the floor. In fact it is only in recent years that most of the manufacturers have had the grace to move the exhaust pipe as far forward as possible. Certainly that has improved things a little, but there is still a long way to go.
- Excellent visibility from the cabin is a minimum requirement. Generally the American manufacturers are bad at this. They delight in erecting instrument panels at eye level so that the view ahead is totally obscured unless you stand six feet, four in your socks. European aircraft are better designed in this respect.
- The aircraft must be stable. Lateral stability is a lost art to so many of the modern designers, consequently you can study the chart on your lap for a few seconds only to find that the hardware has taken on a 15-degree bank and wandered off heading as a result.

Fast touring on low power. Pierre Robin's DR400/Major 80.

- Handling in all three axes must be good. The ideal harmonization of control forces for ailerons, elevators and the rudder is in 2:4:6 proportion, the rudder being heaviest.
- Very low wing loadings might produce slow approach speeds but you get a rough ride in turbulence, while higher wing loadings improve the ride at the expense of higher stalling speed. But you can have the best of both worlds by having effective flaps. I would like my ideal tourer to have area-increasing (Fowler) flaps. And before the sceptics cry 'think of the price', may I remind them that the Cessna 152 and the Aerospatiale Rallye range both have these flaps, yet they are among the cheapest light planes on the market.
- Cruising speeds should be adequate for serious touring. Certainly they must be higher than we are offered today. Time has long passed for a breakthrough in cruising performance. I have in mind a minimum of 130 kt for lower-powered tourers, 145 kt with slightly bigger engines and at least 170 kt by the time we reach the 200 hp plus, retractable gear stage (see next chapter).
- Careful attention must be given to minimizing drag. It is particularly important that good wheel/undercarriage fairings are provided in fixed-gear aircraft. This is often neglected in mass-produced tourers.
- Low-powered tourers should be capable of flying their full cabin for a with-reserve 500 nm and more advanced ones should offer double that range.
- Airframe structure must be as simple as possible, using the smallest number of parts. New methods of construction should be explored with a view to containing the

alarming price increases that have so damaged light aircraft sales in recent years.

I fully acknowledge that few present-day single-engine tourers can meet all or even some of these requirements. But several of the larger manufacturers have tended to rest on their laurels in the mistaken belief that a new paint job or an extra strip of chrome trim will fool customers that this year's model is an improvement on the previous one. The very nature of aeronautics is expensive. We are stuck with a quality-control procedure known as the release system where every single item used in the manufacture of even the smallest light plane is accompanied by a mass of documentation. Materials and labour are costly, engines are far too expensive (and the nearest thing to steam technology you have ever seen – the car manufacturers are far more progressive than the light aero-engine builders), fuel prices climb faster than the aircraft, and therefore it is essential that the modern light tourer should offer its owner the maximum number of air miles per gallon. Reliability is a minimum requirement.

This class of aircraft is usually bought by pilots of low to average experience who intend to go air touring over modest distances (although some very long trips are often undertaken on 160–200-hp fixed-undercarriage types with complete success).

For better or for worse here are some of the single-engine tourers in current use.

Cessna 172 Skyhawk

A Chevrolet with wings

Background

Before the war Cessna were famed for their radial-engined, high-wing monoplane of very clean design. Their production resumed after the war, when the model 195 was introduced with a 300-hp Jacobs radial. It was a muscular bird for a light plane.

No doubt inspired by thoughts of economy the model 170 went into production in 1948. This all-metal, four-seat, tailwheel design flew on a 145-hp Continental engine of the now almost universally adopted horizontally opposed layout. Over 5,000 had been built when production ceased in 1975. Many more might well have been sold but for the introduction of a nosewheel development, which was given model number 172. Cessna made 173 of them during 1955. Since then yearly production has fluctuated between 1,000 and 2,000 units, and well over 35,000 have been sold – a total that must make it the most prolific light plane of all time.

Early models had an upright fin and rudder, a 145-hp Continental engine and a rear fuselage that extended up to the trailing edge of the wing.

A more fashionable swept fin and rudder were introduced in 1960 and a few years later the rear fuselage was lowered to make room for a wraparound window at the back of the passenger area. This brightened the cabin and provided a fair degree of rearward visibility. The name Skyhawk first appeared in 1961 and in more recent times various improvements have included swapping the 145-hp Continental engine for a 160-hp Lycoming.

Engineering and design features

The airframe is of conventional light-alloy construction. A long dorsal fin extends from behind the rear window to blend with the swept vertical surface and its rudder. A fixed tailplane carries separate elevators, a trim tab being fitted to the right-hand section. There is a separate baggage door let into the left side of the fuselage and up to 120 lb (54 kg) may be stowed in the area provided behind the rear seats.

Although the wheels are covered in nicely designed spats, with little doors giving access to the tyre valves, no attempt is made to provide fairings for the nose strut. This detracts from the appearance of this otherwise graceful tourer and must rob it of several knots during the cruise.

The high wing, which is braced to the fuselage by single, wide-chord struts has an untapered centre section. It carries those excellent Cessna flaps – Fowlers that run on the neatest tracking possible. Two fuel tanks in the centre section have a combined capacity of 36 Imp/43 US gal (163

Cessna 172 Skyhawk. As in all high-wing aircraft, good downward visibility is paid for in blindness during turns.

litres) and an optional 45-Imp/54-US-gal (204-litre) system is available.

The tapered outer panels carry those uninspired flat-plate ailerons which should have been replaced with real ones a long time ago. They do little for the aircraft. The flaps, ailerons, rudder and elevators are simple structures with few internal ribs. They rely on the use of corrugated skins to provide rigidity. Externally the standard of finish is good and the Skyhawk is pleasant to look at.

Cabin and flight deck

Two largish car-type doors are provided, entry from either side of the aircraft being assisted by a small step fixed on each of the main undercarriage legs. Access to and from the cabin is not so convenient as it is with some of the other tourers. The front seats have to be slid forward if the two passengers are to avoid gymnastics while getting to the rear of the cabin, and then they need to be moved back again for the benefit of the pilot(s). It is easy to miss the step while alighting from the aircraft and I have known people to trip over the wheel spats. But, with practice, you learn the best way of getting in and out.

By modern standards the 172 is not overgenerous in cabin width. It measures only 40 in (102 cm) across the front seats, although there is an extra 3 in (8 cm) at shoulder level. The deep instrument panel is set rather high for my taste. It has three rows of instruments, and below it a shallow strip runs the full width of the cabin. From left to right it carries the primer, master switch, key-type ignition/starter switch, a line of circuit breakers and then some rocker switches for pitot heat, navigation lights (position lights in the United States), anti-collision beacon, strobe lights and the landing/taxi lights. In the centre is a plunger-type throttle and the mixture control. To their right is the very convenient 'follow-up' flap switch; to save fumbling, gates are provided at 10, 20, 30 and 40 degrees. There is a fresh-air control and another one for the cabin heater; both systems work well. On the extreme right is a small glove compartment.

Four small rectangular instruments provide the usual engine readouts with an rpm indicator nearby set low on the left side of the instrument panel. Adequate space is provided for avionics on the right-hand half of the panel.

A small console continues down to floor level from the centre of the panel. It houses the elevator trim wheel, its position indicator and, when fitted, the optional rudder trim. There is also a fuel selector and a microphone. Between the pilot's knees is a car-type handbrake, which is pulled out from under the instrument panel and twisted to release. Side pockets are provided for maps but there is no proper stowage for airways manuals. Also, there is no annunciator. I would describe the standard of interior trim as plastic and functional rather than lush and plush.

In the air

While taxiing, the 172 drives like a car. Both nosewheel steering and the brakes are effective. Likewise, the take-

But for its untidy nosewheel strut the Cessna Skyhawk looks elegant in the air.

The large instrument panel is mounted high, partly obscuring the view ahead.

off is completely undramatic. You open the throttle wide, there is little noise, no fuss and an in-built talent to run straight down the runway without any prompting by the pilot. Like a well-trained horse it knows the way. I timed 15 seconds from start of the roll to lift-off and the initial rate of climb for our not-fully-loaded Skyhawk was 800 ft/min. Cessna claim 770 ft/min at maximum weight – acceptable if not exactly virile for a 160-hp tourer. At 3,500 ft the 75 per cent power setting (2,500 rpm) returned a TAS of 115 kt. At 8,000 ft this would have become 122 kt. Noise levels are lower than average for a single-engine light plane.

From the pilot's point of view the biggest weakness with the 172 series has always been handling. The elevators are firm but the ailerons and rudder have a soft, spongy feel reminiscent of the more indifferent family cars produced in the 1950s. Possibly Cessna have devised low-geared ailerons to make the 172 drive like an automobile rather than fly like an aircraft.

Flaps-up the Skyhawk stalled at an indicated 42 kt, flap 10 degrees reduced this to 35 kt and with full-flap the airspeed literally went off the clock. There is a lot of position error at high angles of attack and to arrive at the truth you should add about 10 kt to these speeds. Nevertheless the stall behaviour of the 172 makes a canary look like a bird of prey.

In level flight, visibility from the aircraft is satisfactory in most directions, provided the optional height-adjusting seat is fitted (pilots of average height need this to see over the instrument panel). But like most high-wing designs the 172 is very blind towards the centre of a turn.

On base leg and the approach the flaps may be lowered in 10-degree stages as required and I thought Cessna had managed to reduce the trim changes that some pilots found to be a little excessive on earlier models. 70 kt is comfortable initially and the last movement of the flaps from 30 to 40 degrees conveniently reduces the final approach speed to 60 kt without having to retrim. At this speed the aircraft will immediately sink to the runway on all three wheels when the throttle is closed unless more-or-less simultaneous back pressure is applied on the control wheel to ensure correct arrival, mainwheels first.

Although handling is not the Skyhawk's best feature, it is nevertheless very docile and easy to fly.

Capabilities

Ramp weight is 2,307 lb (1,047 kg) and a moderately equipped version would have an empty weight of 1,432 lb (650 kg). Useful load is 875 lb (400 kg) and if your Skyhawk had the optional long-range fuel system, full tanks would mop up 324 lb (150 kg) of that, leaving 551 lb (250 kg) for payload – enough for three adults and an amount of baggage that would vary according to the weight of the occupants. For example, three 170-lb (77-kg) people could share 40 lb (18 kg) of cases, but if their average weight was only 144 lb (65 kg) the full 120-lb (54-kg) baggage allowance could be carried. Thus loaded, it would have a with-reserve range of 630 nm at 75 per cent power (122 kt at 8,000 ft).

Playing it the other way, if you fill the cabin with four 170-lb (77-kg) occupants and 120 lb (54 kg) of baggage, that would represent a payload of 800 lb (363 kg), leaving only 75 lb (34 kg) for fuel, 10.5 Imp/12.5 US gal (47 litres), and with so little in the tanks you should think in terms of a with-reserve range of no more than 100 nm. Of course when the four inmates are less well nourished and the baggage allowance is reduced you can safely fly for greater distances, 300 nm being a representative figure.

Verdict

Why has the 172 been such a success? It does not perform all that well even now that it enjoys an extra 15 hp, its handling is not calculated to inspire the jaded among us, and the cabin is a little on the narrow side. No doubt the boys and girls at Cessna would remind me of the 35,000 that have been sold. They would be entitled to say: 'With a sales record like that who are you to tell us?' (When some critic wrote in less than flattering terms about Liberace, the very successful American entertainer, he is reported to have said: 'I cried all the way to the bank.')

Commercial success is not the only consideration in life and I believe that in this day and age we are entitled to expect better all-round performance and standards of comfort than those provided by the Skyhawk. I know that it is docile, reliable and easy to fly. But does it have to be so *dull*?

Facts and figures

Dimensions

Wing span	35 ft, 10 in
Wing area	174 sq ft
Length	26 ft, 11 in
Height	8 ft, 9 in

Weights & loadings

Max ramp	2,307 lb (1,047 kg)
Max take-off	2,300 lb (1,043 kg)
Equipped empty	1,432 lb (650 kg)
Useful load	875 lb (397 kg)
Seating capacity	4
Max baggage	120 lb (54 kg)
Max fuel Standard:	36 Imp/43 US gal (163 litres)
Optional:	45 Imp/54 US gal (204 litres)
Wing loading	13.2 lb/sq ft
Power loading	14.4 lb/hp

Performance

Max speed at sea level	125 kt
75% power cruise	122 kt
Range at 75% power (optional fuel system)	630 nm
Rate of climb at sea level	770 ft/min
Take-off distance over 50 ft	1,440 ft
Landing distance over 50 ft	1,250 ft
Service ceiling	14,200 ft

Engine

Lycoming 0-320-H2AD producing 160 hp at 2,700 rpm, driving a 75-in (191-cm) diameter propeller.

TBO

2,000 hours.

Piper PA-28-161 Warrior II

Little performance but a lot of comfort

Background

Not so many years ago Piper were building steel-tube, fabric-covered light planes like the Colt and the Tri-Pacer. Then in 1962 John Thorpe designed their PA-28 Cherokee. It was offered in engines of 150, 160, 180 and 235 hp and despite mundane handling and uninspiring performance the type offered a reasonable degree of comfort and it was built like a tank.

In 1974 Piper decided to improve their Cherokee. It had a low-aspect-ratio, slab wing of poor efficiency and the first step was to replace it with tapered outer panels. Of course the introduction of a relatively high-aspect-ratio, tapered wing improved the performance of the tarted-up Cherokee – but then we all knew it would, didn't we (although to hear the way Piper's ad-men talked about it you would have thought they had hit on something new). Engine power was increased slightly from 150 to 160 hp and they called their new plane the Warrior.

Engineering and design features

The wing has no taper over the centre portion of its span, which at 35 ft is 5 ft wider than the Cherokee. However, much of the old Cherokee remains evident in the Warrior. The usual Piper manually-operated, slotted flaps are there along with the usual Cherokee settings: 10, 25 and 40 degrees. The original Warrior had Frise ailerons but later models took a backward step to those flat-plate, piano-wire-hinged affairs which do nothing for the discriminating pilot and rarely provide satisfactory lateral control. Although span has been increased, wing area (170 sq ft) remains unchanged from the original Cherokee. There are two fuel tanks, one in the leading edge of each wing and total usable capacity is 40 Imp/48 US gal (182 litres).

The all-flying tailplane (stabilator) is almost 13 ft in span but only 2 ft in chord. It has an unusually large combined antibalance/trim tab. A dorsal area starts from behind the cabin roof and blends nicely with the modera-

Piper Warrior II, a direct competitor to the Cessna Skyhawk (see previous test).

The Warrior II's instrument panel is a prime example of how to make the most of the available space.

tely swept fin. All control surfaces have fluted skins to increase rigidity.

The fuselage is more or less identical with the Cherokee's. Three windows are set in each side of the cabin and the windscreen has a divider down the centre. Another relic from previous models are the quick-release catches on the engine cowling, which hinges open like an old-fashioned car bonnet (hood). Telescopic struts are used for all three undercarriage legs. The main ones are nicely faired but the nose strut is an untidy creation which reveals bits and pieces that are best hidden from the airflow.

A generously sized baggage compartment is located behind the rear seats, which is reached by a sensible door measuring 20 × 22 in (51 × 56 cm) and up to 200 lb (91 kg) may be carried subject to centre of gravity limitations. Although the Warrior looks what it is, an updated Cherokee, from some angles it is not without grace.

Cabin and flight deck

Only one door is provided, on the right-hand side, and there is a single walkway on the adjacent wing root. To enter you first rotate a roof-mounted handle and then the main release lever is free to open the door. I liked the way it shut without having to slam. You close it gently, then move down a large lever which works within a recess built into the trim, an action which pulls the door tightly into its weather seals.

At 41½ in (105.5 cm) across, the cabin is fractionally narrower than the minimum I suggested at the start of this chapter, although it does offer a little more breathing space than the Cessna 172 (see previous report). For a relatively modest extra charge Piper will do you their special deluxe interior. The example offered me for this airtest had this option and I would regard it as money well spent. With tasteful velour seat covers and simulated black leather trim around the instrument panel glareshield it looked like the interior of an expensive corporate aircraft. Unlike the Cherokee, the Warrior is a proper four-seater, not a '2 plus 2'. There is a bench seat at the back of the cabin for two adults and individual seats for the pilot(s) which may be adjusted for leg reach. Optional height adjustment is also available. The wing has a laminar flow section with maximum thickness occurring some distance back from the leading edge. It is obviously advantageous to position the main spar at the deepest part of the wing and as a result it has been possible to run the spar under the rear seats and thus provide a flat, obstruction-free cabin floor.

Low down on the left-hand wall is a fuel selector which may be set to LEFT, RIGHT and after releasing a safety button, OFF. Rudder trim is adjusted on a small knob under the instrument panel that works by altering the tension of bungee cords. The parking brake lever, which may also be used for applying both brakes simultaneously, hangs down from the centre of the panel and it is locked ON with a small button. There are also toe brakes.

Between the seats a small console carries an elevator trim wheel and the long flap lever. The throttle and mixture levers move within a neat quadrant. The left half of the instrument panel carries flight instruments and radionav readouts. Below them is a strip of five rectangular dials giving fuel contents, oil pressure, cylinder head temperature and electric charge.

To the right of centre is the avionics stack with a row of rocker switches below it for the electric services. A full-cabin-width strip running below the main instrument panel holds the ignition/starter switch, an rpm indicator and all the circuit breakers. The plane tested also had a very neat carburettor ice warning device. Before starting you switch on, adjust a knob to illuminate a blue light, then turn it back until the light extinguishes. You are then correctly set up for the prevailing conditions and the light will warn of any ice build-up in the carburettor. Well-designed control yokes have provision for electric trim and an excellent electronic timer which attaches to the hub. Generally I thought the interior of the Warrior was well planned and most comfortable. However, although two map pockets are provided there is no stowage for manuals. The annunciator lights have a 'press-to-test' button. As in most American light aircraft heating and ventilation are good.

In the air

Starting procedure follows that of any other carburettor-type Lycoming engine. The rudder pedals are linked by springs to the nosewheel and steering is pleasant. By American light plane standards visibility while taxiing is above average. Take-off is normally made flaps-up but for short-field operations 25 degrees is used. You open the throttle and at about 50 kt ease back on the wheel to lift off. It is as simple as that. Initial climb at 70 kt was just over 700 ft/min and noise level was lower than many other light aircraft during this rather vocal phase of flight.

Power settings against flight levels are conveniently printed on the back of the captain's sun visor. At 6,000 ft, 75 per cent power (2,635 rpm) resulted in a TAS of 125 kt, and 65 per cent (2,495 rpm) returned 114 kt. In each case the aircraft was commendably quiet. Visibility, though not up to European standards, is satisfactory.

The stall is totally without vice. Clean, it gently nodded the nose at 53 kt and full-flap will reduce that by 8 kt. There is no wing drop, flaps up or down. When some years ago I flew the original Warrior with its up-market ailerons I thought it was a nice little tourer. The latest model does not handle as well, although I rate it rather better than the Cessna Skyhawk in this respect. Lateral stability is quite good for a modern aircraft but it tends to hunt a little in pitch – not ideal in a touring aircraft. Two cycles were needed to regain a trimmed 110 kt following a 15-kt displacement. I hasten to add that the Warrior is not difficult to fly but Piper should look at means of improving damping in pitch.

Recommended initial approach is 70 kt, going for 63 kt on short finals. Flap limiting speed is a useful 103 kt, by the way. The nosewheel is slightly smaller than the mainwheels, so to ensure adequate propeller clearance you must land the aircraft correctly – not drive it onto the runway in the level attitude. Provided you arrive main-wheels first and then lower the nosewheel, landing a Warrior is a piece of cake.

Capabilities

Maximum take-off weight is 2,325 lb (1,055 kg). A typically equipped example, with a good radio fit and proper flight panel, would have an empty weight of 1,423 lb (646 kg), leaving a useful load of 902 lb (409 kg). So what can we do with that? Well, maximum fuel would take 288 lb (131 kg), leaving a payload of 614 lb (279 kg). This would allow for three 170-lb (77-kg) occupants and 104 lb (47 kg) of baggage. Alternatively you could take two 160-lb (73-kg) men and two 126-lb (57-kg) women, along with 42 lb (15 kg) of luggage. With full tanks you can fly a with-reserve 590 nm at 123 kt (75 per cent power) or 633 nm when the 65 per cent setting has been used to produce a TAS of 116 kt. Playing it the other way, four 160-lb (73-kg) adults and 160 lb (73 kg) of baggage would mean limiting fuel to 15 Imp/18 US gal (68 litres) and at 75 per cent power range would be 180 nm.

Verdict

It is no secret that Piper introduced the Warrior to compete with the very successful Skyhawk. In terms of performance there is little to choose between the two aircraft, but the Warrior is slightly wider inside and, to me at any rate, more attractively turned out. Whether or not you prefer high-or low-wing designs is a matter of personal taste. I am a low-wing fan myself, particularly since my local airfield is slacking if there are not more than eight aircraft on the circuit – and I like to see the bandits even while turning.

I think it is a pity that Piper have opted for stabilators in

The slightly tapered wing, replacing the slab mainplane of the Cherokee, has made the Warrior a proper four-seat tourer.

so many of their aircraft. They originated when supersonic fighters came on the scene with their wide speed ranges and correspondingly wide centre-of-pressure movements which demanded a great deal of pitch control authority. But fast fighters have powered controls, while light planes still rely on good, old-fashioned 'push-and-grunt'. And I have yet to fly any light single or twin with a stabilator which matches the quality of handling provided by a fixed tailplane (stabilizer in the US) and separate elevators.

In terms of engineering the Warrior is a very robust airframe and the type has a good maintenance record at some of the world's largest flying schools. However, at time of writing it is coming up eighty years since the Wright's Flyer became the first powered fixed-wing aircraft to get itself airborne. And I think by now we should be entitled to a little more payload/range and cruising performance on a 160-hp engine than is provided by either the Cessna Skyhawk or the Piper Warrior. After all, neither of them has a particularly wide cabin. Having said that, the Warrior is nevertheless a comfortable and predictable low-powered tourer. I think I could grow to like it a lot,

Facts and figures

Dimensions

Wing span	35 ft
Wing area	170 sq ft
Length	23 ft, 9 in
Height	7 ft, 4 in

Weights & loadings

Max take-off	2,325 lb (1,055 kg)
Equipped empty	1,423 lb (646 kg)
Useful load	902 lb (409 kg)
Seating capacity	4
Max baggage	200 lb (91 kg)
Max fuel	40 Imp/48 US gal (182 litres)
Wing loading	13.7 lb/sq ft
Power loading	14.5 lb/hp

Performance

Max speed at sea level	127 kt
75% power cruise	123 kt
65% power cruise	116 kt
Range at 75% power	590 nm
Range at 65% power	633 nm
Rate of climb at sea level	710 ft/min
Take-off distance over 50 ft	1,490 ft
Landing distance over 50 ft	1,115 ft
Service ceiling	13,000 ft

Engine
Lycoming 0-320-D3G developing 160 hp at 2,700 rpm, driving a 74-in (188-cm) diameter Sensenich propeller.

TBO
2,000 hours.

Robin DR400/Major 80

Sports car of the air

Background

In my airtest of the Robin DR400/Dauphin in Chapter 2 I explained the background to the 400 series and mentioned that the same airframe is offered with engines of 120, 160 and 180 hp. The Major 80 represents the middle of the Robin DR400 range and my subtitle, 'Sports car of the air' is, I believe, a fair description of this remarkable tourer.

The lower-powered Dauphin is capable of cruising four people faster than two-seat trainers flying on the same engine. The Major 80 takes this outstanding performance a stage further in offering another 300 nm range while cruising some 20 kt faster. There is also a 230 ft/min improvement in climb rate – all for an increase of only 40 hp.

It is common knowledge that every airframe has an optimum engine size. Reduce power below what is ideal and performance suffers; increase engine power above optimum and there will be a reduction in range (unless the fuel capacity is enlarged), with little improvement in cruising speed – although rate of climb will benefit from the surplus power. I would imagine that 160 hp is ideal for the DR400 airframe because it offers the significant performance improvements over the 120-hp Dauphin I have already mentioned, while its 180-hp brother does little better in terms of cruise performance. In fact you would be hard pressed to find another 160-hp aircraft that can match the overall efficiency of the Major 80.

Engineering and design features

The same Jodel-type wing as used for the Dauphin forms the basis of the DR400/Major 80 with the exception that two small fuel tanks, each of 8.8 Imp/10.5 US gal (40 litres) capacity are built into the wing-root leading edges. These supplement the main fuel tank, which resides under the baggage area behind the rear seats. Total usable fuel of all three tanks is 40 Imp/48 US gal (182 litres).

The same rather crude ailerons and simple flaps are used and the highly tapered outer wing panels carry 14 degrees of dihedral. They are given 5 or 6 degrees of washout. This is very noticeable when the aircraft is viewed in side elevation, and during cruising flight almost half of the wing is presented at a low-drag/low-lift angle of attack, most of the aircraft being supported by the untapered centre section.

The leading edges of the wing sweep forward as they meet the fuselage and provide a useful location for the two wing tanks. A filler cap is conveniently let into each leading edge, the one for the main tank being situated behind the cabin, on the left side of the fuselage.

The fuselage is an excellent blend of aesthetics, structural

Note the efficient Jodel cranked wing and the large canopy that slides forward when opened.

practicality and aerodynamic realism. It is a strong wooden box, rounded in the right places to reduce drag but not to the point where cabin comfort suffers; 43.5 in (110.5 cm) wide, 48.5 in (123 cm) high and 64 in (162.5 cm) long are measurements that are nothing to be ashamed of in a

low-powered tourer.

The rear fuselage blends smoothly into a swept fin and the stabilator has antibalance tabs added to its trailing edge, their datum being altered for trim purposes on a wheel positioned between the two front seats. Pierre Robin's fixed undercarriages are the best in the business. The nosewheel strut is a model of how it should be done, with none of your 'torque-links-dangling-in-the-breeze' nonsense. After take-off when the weight is removed from the wheels, the nose strut automatically disconnects from the rudder pedal steering and aligns itself neatly down the fore and aft axis of the aircraft.

To climb aboard the lugger, an act that may be performed on either side, you step onto the wing walkways, twist a roof-mounted handle through 90 degrees and then slide the canopy and its all-in-one windscreen forward to reveal most of the cabin area. The two front seats tip forward to make life easy for the pair of inmates

The instrument panel sits below the line of sight.
There is an annunciator strip in the glareshield and avionics
may be fitted in the space next to the right-hand throttle.

destined for the rear bench. But before entering the aircraft, a word about that very tidy engine installation up front. The fibreglass cowlings are beautifully designed and the carburettor air intake protrudes forward, where it takes advantage of ram pressure. Never has a 160-hp Lycoming engine been more elegantly clothed.

Cabin and flight deck

The glareshield carries an annunciated strip with the usual warning lights and a 'press-to-test' button. In front of the captain's seat is the flight panel with the radionav readouts on its right. An area to the right of centre will accept a comprehensive radio stack and the rpm indicator is on the right-hand panel.

Along the bottom of the main instrument panel is a strip running the full width of the cabin. It contains the master switch, key-type ignition/starter, switches for the electric services which have a thermal overload facility to protect the circuit. (You press a green button for, say, the pitot heat and a red one switches it off.) There is a bank of small engine instrument and fuel gauges, the carburettor heat and mixture controls and, on the extreme right, the effective cabin heat/fresh-air selector knobs.

Two plunger-type throttles are provided, one on the left and another in the centre, which have no separate friction adjustment, an arrangement that, frankly, I do not care for.

A narrow console extends down from the centre of the instrument panel and back between the front seats. It accommodates a large and convenient-to-use brake handle which brings on both brakes together and locks them on for parking by twisting through 90 degrees. While taxiing, nosewheel steering via the rudder pedals is supplemented by differential braking brought on when the rudder pedal approaches the end of its travel in each direction. Toe brakes are available as an optional extra. There is a convenient trim wheel with a large position indicator and a lever for the flaps. These may be set to UP, 15 degrees and 60 degrees. Two control sticks are provided which bend their way under the instrument panel to cause as little inconvenience to the pilot's legs as possible. There are large map pockets which at a pinch will take airways manuals.

The front seats offer good support. They may be adjusted through six positions moving up as they slide forward. The cabin is workmanlike, functional, easy to live with, but not plush.

In the air

The first impression if you are not previously acquainted

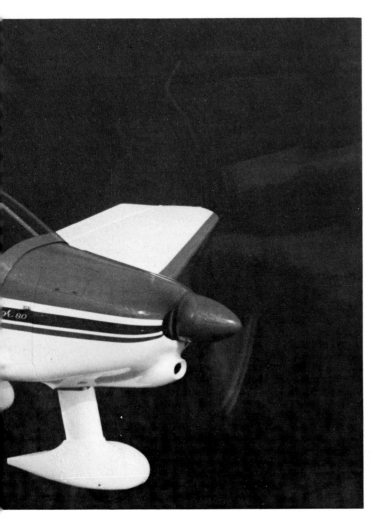

Note the deep windows and the neatly faired nosewheel.
Compare this with the Cessna and Piper light singles.

with Robin aircraft is of the quite astonishing visibility. You can see ahead over the nose, and the massive side windows extend down to elbow level providing helicopter-like views.

Nosewheel steering is good and the disc brakes are particularly powerful. The test aircraft was about 300 lb (136 kg) below its maximum take-off weight so we tended to leap off the ground after rotating at 55 kt. However, I have since flown fully loaded examples and there is little difference in take-off performance. Initial rate of climb was 1,200 ft/min, which was more than 200 ft/min better than claimed for the Major 80 at maximum weight. With 75 per cent power you readily see 125 kt indicated but at 8,000 ft you can expect a TAS of 137 kt with a full load – outstanding for a low-powered tourer providing its occupants with a generous cabin. At that speed, endurance is quoted as almost five and a half hours.

All controls are rather firm but the ailerons are effective and the aircraft has a live, responsive feel which, allied to its outstanding visibility in most directions, make it a great little fighter to fly. Stability is good in all three axes and noise levels are slightly quieter than average, but could still do with improvement.

The clean stall occurs at 50 kt and the rather ineffective flaps take off 5 kt at the 60-degree maximum setting.

On the approach, visibility is outstanding, there are few trim changes and 65 kt is ideal for short finals. Landing could not be simpler. The Major 80 certainly puts the fun back into flying.

Capabilities

Maximum weight is 2,315 lb (1,050 kg) and a typically equipped example would have a useful load of 975 lb (442 kg) which is, in round figures, 75 lb (34 kg) more than a Piper Warrior (see previous test report). Full tanks will allow a payload of 687 lb (312 kg) – enough for four 170-lb (77-kg) adults and about 7 lb (3 kg) for odds and ends. On the other hand, four 160-lb (73-kg) occupants could take with them about 50 lb (23 kg) of baggage. Full tanks will fly a with-reserve 650 nm at 137 kt (75 per cent power) or 750 nm at 131 kt – both are excellent figures. Maximum baggage, however, is only 90 lb (41 kg), not enough for a four-seat tourer in my opinion. Alternatively, a full cabin (four occupants each of 170 lb (77 kg) plus 90 lb (41 kg) of baggage) would entail reducing fuel by 11.5 Imp/14 US gal (53 litres) when the ranges become 470 nm and 540 nm respectively. When you consider that the full cabin range of a Cessna Skyhawk is only 100 nm or so the capabilities of this remarkable little aircraft can be fully appreciated.

Verdict

One of the reasons for the Major 80's outstanding overall performance lies in its very clean airframe, the result of painstaking attention to detail. Pierre Robin fights a constant battle with drag and this has endowed the Major 80 with a power-off glide angle of 1 in 9.8, representing a lift/drag ratio of almost 10.

There is a lot of ill-informed prejudice against wooden aircraft. Modern materials and synthetic glues have transformed such airframes into durable, fatigue- and corrosion-free structures and you have only to see the condition of ageing DR400 aircraft, many of them taking a beating at the flying schools, to appreciate their hard-wearing qualities.

In terms of performance the DR400/Major 80 is 14 kt faster at 75 per cent power than the Cessna Skyhawk or the Piper Warrior, its rate of climb is 120 ft/min faster, its payload range is in a different class and it handles better than the two other aircraft. The last comparison is a matter of personal opinion but the others are facts based upon performance figures.

Being so fast for its power the Major 80 is very economical, and the standard of construction and finish is first class throughout. Pierre Robin's Major 80 is truly a sports car of the air.

Facts and figures

Dimensions

Wing span	28 ft, 8 in
Wing area	153 sq ft
Length	22 ft, 10 in
Height	7 ft, 4 in

Weights & loadings

Max take-off	2,315 lb (1,050 kg)
Equipped empty	1,340 lb (608 kg)
Useful load	975 lb (442 kg)
Seating capacity	4
Max baggage	90 lb (41 kg)
Max fuel (usable)	40 Imp/48 US gal (182 litres)
Wing loading	15.13 lb/sq ft
Power loading	14.47 lb/hp

Performance

Max speed at sea level	146 kt
75% power cruise	137 kt
65% power cruise	131 kt
Range at 75% power	650 nm
Range at 65% power	750 nm
Rate of climb at sea level	830 ft/min
Take-off distance over 50 ft	1,640 ft
Landing distance over 50 ft	1,780 ft
Service ceiling	13,500 ft

Engine
Lycoming 0-320-D developing 160 hp at 2,700 rpm, driving a 72-in (183-cm) or 74-in (188-cm) Sensenich propeller.

TBO
2,000 hours.

Robin R3140

An all-metal Jodel

Background

A brief account of the factors that induced Dijon light plane manufacturers Avions Pierre Robin to design and build metal aircraft was given in the Robin R2160 airtest report in Chapter 2. As already mentioned, the first metal design was the HR100, a high-quality, four-seat tourer with four fuel tanks in the wings holding a total of 100 Imp/120 US gal (454 litres). The wing may have been a good fuel tank but it did little for the HR100's performance other than give it a phenomenal 1,500 nm range.

In 1978 there followed a lower-powered, cheaper-to-build tourer called the R118OT Aiglon which, like the previous metal Robins, had a parallel-chord slab wing. In terms of cruising speed the Aiglon was on a par with the best American 180-hp fixed-gear tourers although it offered a wider cabin, better visibility and more useful load. However, Pierre Robin was forced to accept that in entering the tin airframe business he had made the mistake of ditching that all-conquering, unbeatable, Jodel wing. So although the Aiglon was a relatively new design with a lot of development potential Pierre got himself a clean sheet of paper, a new battery for his pocket calculator and sharp pencils.

A new breed of light planes emerged called the R3000 series. It was first shown at the 1981 Paris Air Show. The intention is to offer two versions of the same basic airframe. The Panoramic version has a typically European canopy over the cabin area giving outstanding views in

The prototype Robin R3140.

Auto-steps are built into each flap to assist entry and exit.

most directions, while the Integral Fuselage models will be more like an American light plane with windows let into the side like an automobile. Both versions will be marketed with a choice of engines ranging from 100 hp (trainer version) up to 200 hp. Turbocharged retractables of 180 and 200 hp are also planned, but the subject of this report is the R3140. At time of publication this interesting aircraft is a prototype powered by a 160-hp engine – so really Pierre Robin should call it the R3160.

Engineering and design features
You may recall that the Jodel wing features a large parallel-chord centre section without dihedral, to which are added highly tapered outer panels. These enjoy 14 degrees of dihedral and about 8 degrees of washout so that in cruising flight most of the lift (and drag) is confined to the centre section, the tapered areas only coming into action at lower airspeeds.

The metal wing of the R3140 has constant dihedral from root to tip but follows the Jodel planform so the outer panels taper in chord and thickness. Spar depth must reduce in proportion to the chord of the wing and all taper is on the under-surface of the outer panels, in effect increasing their dihedral angle. There is also considerable washout and the tips are fitted with up-swept winglets to combat induced drag and spanwise flow during the cruise.

Well-made Frise ailerons and slotted flaps replace the somewhat crude flat surfaces fitted to the wooden DR400 series, and the inboard ends of the flaps are provided with the neatest imaginable autostep for assisting entry and exit. As the flap is lowered, a small area near the wing root

The well-planned instrument panel includes large shelves for maps and airways manuals.

hinges up parallel with the ground to provide a convenient step.

The prototype has two wing tanks with a total usable capacity of 30 Imp/36 US gal (136 litres). That, in my view, is inadequate but a 50-Imp/60-US-gal (227-litre) system is contemplated for production aircraft.

The fuselage is very simply constructed with few frames and stringers, semiflush rivets being used in most areas. A large dorsal fin extends from behind the cabin to the main swept area which carries a fixed tailplane and separate elevators on top. Whatever fans of these 'T' tails may claim of them (and I suspect many of the so-called advantages are more imaginary than real) the one they have given to the R3140 does nothing for its appearance.

Like all Robin aircraft the air intake is projected well forward of stagnant or turbulent air, where it can gain full advantage of ram effect and so add a little power. On the other side of the equation, the three undercarriage legs and their wheels are enclosed in carefully designed fairings aimed at reducing drag to a minimum. All parts of the airframe are given proper anti-corrosive treatment *before* assembly. This is standard practice at Avions Robin – not an expensive option.

Seen on the ground, the R3140, with its massive side windows and clean lines, is a graceful light aircraft, its beauty of line being marred slightly by that 'T' tail, which looks to me as though it belongs to another aircraft.

Cabin and flight deck

You mount either the left- or right-hand wing walkway using the excellent autostep previously mentioned, twist the roof-mounted handle and slide forward the one-piece windshield and canopy which surrounds the front seats. Passengers sitting in the back of the cabin gain their places by tilting forward the front seats and they can enjoy the unique windows that start at elbow level and provide such a magnificent view. Entry and exit is easier than for most small aircraft.

From side to side the cabin is about 45 in (114 cm) wide at shoulder level. There is ample room for four people and a baggage area is provided behind the rear seats. The prototype has no external baggage door but I understand this is being considered for production aircraft. The present 88-lb (40-kg) limit should be increased to 120–130 lb (55–60 kg). Four people are entitled to carry a reason-

Clean lines allied to a Jodel-type wing result in an outstanding low-powered tourer.

able amount of luggage; 22 lb (10 kg) a head is only half the miserable airline allowance.

A small panel built into the left sidewall carries the battery master switch and others for the various electric services. Plenty of room is available for radio on the right-hand panel with flight instruments and radionav readouts in front of the left-hand seat. The usual cabin vent and heat controls are provided.

Extending down from the main instrument panel and running back between the front seats is a central console. From the top running down and back are the engine instruments followed by the throttle, mixture control and carburettor heat. An adjustable friction nut is on the right-hand side of the console. The toe brakes are locked for parking on a small handle below the engine controls, then there is a splendidly designed elevator trim wheel and a good trim-tab position indicator. Nearby is a spring-loaded switch for the flaps. Their position is indicated on three green lights included on the annunciator strip built into the glare-shield. The lights indicate settings of 10, 20 and 30 degrees. The console also carries the fuel contents gauges and the tank selector.

On either side of the central console are two deep shelves providing convenient storage space for flight manuals, maps and other navigation equipment. From the centre of each deep padded shelf emerge the push-pull/torque tubes and their control yokes.

Viewed in its entirety the cabin looks like the interior of an expensive GT car. The Robin outfit has always been good at flight deck design. This is one of their best efforts.

In the air

The prototype had no rear seats but there was lead ballast and full fuel to bring us up to maximum weight. Starting is as usual for a carburettor-type engine. The nosewheel is linked by springs to the rudder pedals, which on the prototype were a little soft. For my liking there was not enough steering authority, but you have to make allowances for a prototype under development.

Flap 10 degrees is used for take-off. The little airfield adjacent to Avions Pierre Robin is rough and almost 2,000 ft above sea level. It was also a very hot day yet we went up at 900 ft/min – my first confirmation of the superior qualities of Jodel-type wings.

At 4,000 ft, 75 per cent power (2,550 rpm) resulted in a TAS of 120 kt, but normally aspirated aircraft are at their best flying between 7,000 and 12,000 ft. At 8,500 ft, for example, 75 per cent power will provide a 133-kt cruise, and at 12,000 ft the economical 65 per cent setting will fly a full load at 120 kt. These speeds are equivalent to the better 180-hp tourers, so, in practical terms, the new wing is probably worth 20 hp.

Visibility is in true Robin tradition – superb. It was difficult to assess noise levels properly because there were no seats in the rear and upholstery tends to absorb sound. But flown in prototype form I would describe the cabin as a little quieter than average – not among the most peaceful.

Handling, even at this stage of the plane's development, shows great promise. The ailerons and rudder are first class, but there is some sponginess in the elevator circuit caused by flexing in the coupling between the two control wheels. Lateral stability is unusually high for a modern aircraft; and what a boon that is while doing a one-arm juggling act on the maps or radionav charts! Pitch stability is outstanding: it corrected a hands-off 20-kt displacement from its trimmed speed in only one cycle – a really tough test. Directionally there is a slight tendency to hunt left and right, which many pilots would not even notice. But the Robin team (and yours truly) have and the problem had been dealt with before this book went into print.

Stalling speeds are remarkably low but this is deliberate policy. The company intends to market a 100-hp, two-seat trainer version of the airframe. Flaps-up the 'g' break came at 50 kt and full-flap (30 degrees) brought the stall down to only 40 kt. During the approach, trim changes are slight until the final stage of flap is applied, but this too is being dealt with by using a simple mechanical linkage between the flaps and the trim tab. Over-the-fence speed is 65 kt, the elevators retaining their power until speed has been reduced to a modest trot. The landing is simple, straightforward and predictable.

Capabilities

Maximum take-off weight is 2,205 lb (1,000 kg), 100 lb (45 kg) less than a Cessna Skyhawk and 120 lb (54 kg) lighter than a Piper Warrior. Yet its 885-lb (401-kg) equipped useful load exceeds that of the former and almost equals the latter's. When four 160-lb (73-kg) adults are carried along with 88 lb (40 kg) of baggage there is enough fuel, 22 Imp/26 US gal (98 litres), for a with-reserve range of about 360 nm. A Piper Warrior will carry the same load for 300 nm and the Cessna Skyhawk will manage 225 nm. In each case the Robin R3140 has a 10-kt speed advantage at 75 per cent power.

Full fuel, assuming the larger system, amounts to 360 lb (163 kg) and that would allow 525 lb (238 kt) to be placed in the cabin – enough for three large adults and a little baggage. Such a load could be flown for 830 nm, which is outstanding for a 160-hp tourer. In terms of field performance the Robin can match the Piper or the Cessna, while its rate of climb is far superior.

Verdict

It is early days for the Robin R3000 series, so while the prototype R3140 is a little slower than the wooden DR400/

Major 80, the latter's design has many years of development behind it. The R3140 has better field performance than its wooden ancestor and a faster rate of climb. It can also out-range the Major 80 and provide a more comfortable cabin.

Compared with aircraft from the other manufacturers, the R3140 shines as a good all-rounder, more or less equalling the performance of 180-hp fixed undercarriage tourers but with greater fuel economy.

The R3000 range also includes some exciting higher-powered versions which are in the pipeline. For example, there is a projected 180-hp retractable that will cruise at 152 kt and a turbocharged version which will offer 178 kt at 75 per cent power. But I have ceased to be surprised at anything that emerges from the little Dijon factory in the heart of France.

Facts and figures

Dimensions

Wing span	32 ft, 2 in
Wing area	156 sq ft
Length	24 ft, 7 in
Height	8 ft, 9 in

Weights & loadings

Max take-off	2,205 lb (1,000 kg)
Equipped empty	1,320 lb (599 kg)
Useful load	885 lb (401 kg)
Seating capacity	4
Max baggage	88 lb (40 kg)
Max fuel Standard	30 Imp/36 US gal (136 litres)
Optional	50 Imp/60 US gal (227 litres)
Wing loading	14.13 lb/sq ft
Power loading	13.78 lb/hp

Performance

75% power cruise at 8,500 ft	133 kt
65% power cruise at 12,000 ft	120 kt
Range at 75% power (optional tanks)	750 nm
Range at 65% power (optional tanks)	830 nm
Rate of climb	900 ft/min
Take-off distance over 50 ft	1,445 ft
Landing distance over 50 ft	1,150 ft
Service ceiling	14,500 ft

Engine
Lycoming 0-320-D2A developing 160 hp at 2,700 rpm, driving a Sensenich two-blade, fixed-pitch propeller.

TBO
2,000 hours.

Aerospatiale TB10 Tobago

New standards in cabin space

Background

I have given a brief run-down on light-aircraft production at Aerospatiale, the giant French aerospace company, in my airtest report on the Rallye Galopin in Chapter 2. There I mentioned that the Rallye, which won the French government's design competition for a safe, cheap trainer, was built with engines of from 90 to 235 hp. The higher-powered Rallyes were intended as four-seat tourers but, as already explained, there was so much drag created by all the various bits of ironmongery that festooned the beslatted, Fowler-flapped wing of the breed that the addition of power had a very limited effect on cruising speed – although, of course, weight, lifting capabilities and rate of climb improved. The other problem was one of comfort, and was caused by the position of the main spar, which, running through the cabin, left little room for your feet while sitting in the back.

The Rallye is still produced for certain specialized tasks (training, crop spraying, glider towing and so forth) but understandably the Aerospatiale concern decided that if it was to compete in the single-engine tourer market, time had arrived for a fresh start with a clean sheet of paper. It was a case of 'back to the drawing board'. As a result, there emerged in 1978 a totally new range of tourers, incorporating a number of radically new features. The basic airframe was designed with stretch capabilities in mind, thus there is a 160-hp, fixed-pitch propeller version (TB9 Tampico); a similar one with a constant speed airscrew (the 180-hp TB10 Tobago); and a 250-hp retractable called the TB20 Trinidad described in the next chapter. For the future a light twin is also contemplated.

Here I shall be concentrating on the 180-hp Tobago, but I have also included performance figures for the lower-powered Tampico in the data panel at the end of this report.

Engineering and design features

The TB9, 10 and 20 range of tourers is built at Aerospatiale's Tarbes factory. Many of the parts come off the same sophisticated machines as those used in the manufacture of Mirage fighters, Falcon bizjets, advanced helicopters and parts of the outstanding Airbus. This is contrary to British and American thinking in terms of production economics but in my view the French are right and we are very wrong. Their attitude is: 'We can't utilize these expensive stretch formers, presses and anti-corrosion plants for every minute of the working day. So rather than have them stand idle (costing a bomb in lost utilization) why not stamp out light plane parts?' Why not indeed!

Aerospatiale TB10 Tobago, a breakaway from tradition in many respects. Within its graceful lines is the widest cabin of any single-engine light plane.

The airframe is based upon five modules, each of them a simple structure with very few detailed parts. When an aircraft is ordered, radio and other options can be installed at an ideal stage before the engineers start getting in one another's way. The wing of only 128 sq ft area (a Cessna Skyhawk's is 174 sq ft) is in complete contrast to the earlier Rallye. There are no autoslats, Fowler flaps are replaced by the slotted type, and yet they have managed to avoid high stalling speeds by using a modern, computer-generated airfoil section. Two wing tanks have a total fuel capacity of 45 Imp/54 US gal (204 litres) usable. The two filler caps are of large diameter and they may be locked.

The instruments and radio are in padded boxes instead of the usual panels.

First impression of the fuselage is of a long, shapely nose. Fibreglass cowlings may easily be removed by undoing a number of quick-release screws. The exhaust pipe is under the front of the nose and quite a long way from the cabin. The wing passes through the fuselage below floor level, which is an obvious advantage from the passengers' point of view. Behind the trailing edge is a pair of strakes mounted under the fuselage for the purpose of improving rudder response and directional stability at low airspeeds.

A very elegant swept fin and rudder is set ahead of the stabilator and – surprise, surprise – they managed to get it right first time, because there is no additional dorsal area. At first examination the stabilator appears to have no down movement, but relative to the wing chord line, of course, it has.

On the left side of the fuselage is a small triangular door leading to the baggage hold. I have suggested to the manufacturers that it is too small for convenience; there are no structural reasons why it should not be enlarged and I understand this is being considered. Likewise I thought the 100-lb (45-kg) baggage allowance was not overgenerous, particularly since you can seat up to five people in these aircraft.

All wheels are nicely faired but the nosewheel strut is on a par with most American fixed-gear light planes. No attempt is made to hide those torque links. Only Pierre Robin knows how to do it properly (see previous test reports).

Seen on the ground the aircraft has a lot of style. Much use is made of flush riveting, and Aerospatiale's claim that their light planes are built to the same standards as their larger products is no flight of fancy. I have seen for myself during a tour of the factory.

Two gull-wing doors are provided, which you must first open before climbing onto the wing. The two front seats are well designed and a conveniently placed catch allows you to tilt their backs forward to allow easy access for up to three passengers in the rear of the cabin.

Cabin and flight deck

On entering the aircraft you immediately recognize that it is quite unlike any other. It could belong to a space movie. Then you become aware of the S – P – A – C – E, lots of it – even in automobile terms. At shoulder level the cabin is 50 in (127 cm) wide – the same as the Piper Navajo, an eight-passenger light twin (see Chapter 5), and there is more leg room than in many a large car. The cabin is something of a breakthrough in light-aircraft design.

Standard seats are covered in a tasteful material, featuring wide bands of contrasting colour. Gone is the traditional instrument panel, to be replaced by boxes padded in simulated leather, an upright one in the centre for the avionics and below them, a line of vertical engine readouts. To the left is a horizontal padded box containing the

flight and radio instruments, the complete section being removable for maintenance by releasing a single screw. If required, a duplicate box of instruments may be installed in front of the right-hand seat, otherwise there is a useful shelf on that side.

A wide console extends back between the pilots and provides a home for the throttle, pitch and mixture levers, an electric flap switch, flap position indicator, trim wheel and indicator and a hand microphone.

A good annunciator is provided and the electrics are controlled on a line of thermal overload buttons which are

Beauty in flight spoiled by an untidy nosewheel strut.

identified by those supposedly international symbols so beloved by the French (they look like Chinese flute music to me). Pockets are provided for maps, airways manuals, and so on, and a convenient flap, complete with built-in clip, lets down from the ceiling where it can be used to hold an approach chart. There is a good heater and fresh air system.

The doors close easily without slamming. You just reach up, hinge them down, then operate the handle, which pulls the door into its weather seals. Massive windows are provided and visibility in most directions can only be described as excellent. Only the Robin aircraft are slightly better in this respect – but now I am splitting hairs.

It would be fair to say that the Tobago cabin represents the first break with tradition in light-aircraft design for many years. In terms of space and grace it is something of a trendsetter.

In the air
At the time of my airtest the aircraft was within 250 lb (113 kg) of its 2,535 lb (1,150 kg) maximum weight. You start the carburettor-type Lycoming like any other and the

aircraft drives as easily as a car while taxiing. Apart from the rather thick door pillars, visibility is excellent, the brakes are powerful and the nosewheel steering provides accurate control of direction.

On opening the throttle you get the impression that the bird will not take off: there is little noise and acceleration is not very brisk. But it is all an illusion because at maximum weight a Tobago requires no more take-off distance to the 50-ft point than a Piper Archer II (the 180-hp version of the Warrior described earlier in this chapter). Lift-off speed is only 65 kt and the 25-kt crosswind limit is higher than that for many of the light twins.

At 3,500 ft, 2,700 rpm/22.1 in manifold pressure (about 75 per cent power for that level) gave an IAS of 125 kt but, believe it or not, I was unable to compute the TAS because this fine, well-equipped tourer did not have an OAT gauge (although many of the bits and pieces included in the standard aircraft are considered optional extras by other manufacturers – full corrosion proofing, to mention one). However, the aircraft manual quotes a range of power settings and at 4,000 ft 75 per cent will return a TAS of 123 kt at maximum weight. Fastest cruise mentioned is 127 kt but a good all-round technique would be 2,300 rpm/22.1 in manifold pressure which at 6,000 ft will fly a with-reserve 560 nm at 121 kt. At this power setting the Tobago is very quiet; personally I thought the aircraft sounded sweeter at 2,700 rpm. This eliminated some of the low-frequency rumble.

The stall is preceded by a melodious bell and without flap came at 55 kt, with 45 kt being indicated in the full-flap configuration.

Turns can be made without need of rudder to maintain balance. The rudder and elevators are light by modern standards but the effective ailerons are too heavy for ideal control force harmonization. Like most modern aircraft there is very little lateral stability but overall stability is good. This reads across to the approach where trim changes are minimal and the threshold speed is only 65 kt. The landing itself is perfectly straightforward and I tried a full-flap emergency take-off, leaving the runway after 13 seconds at only 55 kt.

Capabilities

Equipped empty weight is around 1,500 lb (680 kg), which leaves a useful load of 1,035 lb (470 kg). Full tanks will allow a payload of 709 lb (322 kg), say, four 170-lb (77-kg) adults and 89 lb (40 kg) of baggage. That little lot can be flown for a with-reserve range of 550–580 nm, according to speed and cruising level. You can take three people on the back seat but fuel load would have to be reduced accordingly. So if the Tobago is not the fastest 180-hp tourer on the market it is certainly the one with the biggest cabin.

Verdict

With a cabin 7 in (18 cm) wider than a Piper Warrior/Archer and no less than 9 in (23 cm) wider than the Cessna

Hawk XP, you would expect the Tobago to give away some knots in cruising speed. Actually it does quite well in this respect and its 75 per cent 127 kt is only 4 kt slower than the similarly powered Archer. It is actually faster than the Beech Sundowner. Its equipped useful load is better than the Cessna's and the Beech's, and is within a few pounds of the Piper's. So its magnificent cabin has not been at the expense of all else. And it is something of a design breakthrough.

Facts and figures

Aerospatiale TB9 Tampico		Aerospatiale TB10 Tobago
Dimensions		
32 ft	Wing span	32 ft
128 sq ft	Wing area	128 sq ft
25 ft	Length	25 ft
10 ft, 6 in	Height	10 ft, 6 in
Weights & loadings		
2,330 lb (1,057 kg)	Max take-off	2,535 lb (1,150 kg)
1,450 lb (658 kg)	Equipped empty	1,500 lb (680 kg)
880 lb (400 kg)	Useful load	1,035 lb (470 kg)
4/5	Seating capacity	4/5
88 lb (40 kg)	Max baggage	100 lb (45 kg)
34 Imp/41 US gal (155 litres)	Max fuel	45 Imp/54 US gal (204 litres)
18.2 lb/sq ft	Wing loading	19.76 lb/sq ft
14.56 lb/hp	Power loading	14.05 lb/hp
Performance		
122 kt	Max speed at sea level	133 kt
120 kt	75% power cruise	127 kt
108 kt	65% power cruise	117 kt
370 nm	Range at 75% power	560 nm
660 ft/min	Rate of climb at sea level	800 ft/min
1,853 ft	Take-off distance over 50 ft	1,660 ft
1,330 ft	Landing distance over 50 ft	1,395 ft
12,500 ft	Service ceiling	13,000 ft
Engine		
Lycoming 0-320-D2A developing 160 hp at 2,700 rpm, driving a fixed-pitch propeller.		Lycoming 0-360-A-1AD developing 180 hp at 2,700 rpm, driving a two-blade, constant-speed propeller.
TBO		
2,000 hours.		2,000 hours.

Cessna Turbo Stationair 6

The flying station-wagon

Background

If the Cessna 172 (see page 48) may be called a 'Chevrolet with wings', then without doubt the subject of this airtest has got to be the 'flying station-wagon'. It all started with the model 206, at the time known as the Super Skywagon (Cessna really must do something about those ad men of theirs). Cessna's 206 was introduced in 1965 and a series of improvements led to the Stationair, which came on the market in 1971. A year later they fitted a new wing with an airfoil section claimed to offer improved low-speed handling. As its name implies the Stationair 6 will carry six people. There is also a Stationair 7, which sports a lengthened fuselage and a row of cabin windows like a school bus. Behind the three pairs of seats is a single one for an additional passenger. Both aircraft are offered with a choice of engines, the normally aspirated Continental 10-520-F developing 300 hp or the turbocharged, 310-hp TS10-520-M version, which has the advantage of maintaining power at altitudes where the unsupercharged of the species would be gasping for breath.

Engineering and design features

The familiar Cessna wing used on the Stationair has the same area as the much lighter and lower-powered 172 Skyhawk. It is graced with those magnificent Fowler flaps, and the crude flat-plate ailerons as fitted to little brother are replaced by well-designed Frise control surfaces. Usable fuel capacity is 73 Imp/88 US gal (333 litres) and you have to climb up to reach the filler points which are located on top of the wing.

A fixed tailplane and separate elevators are fitted. In the main, Cessna have resisted the temptation of adopting stabilators which, in my experience, never handle in the pleasant way of elevators. The swept fin has a large dorsal area which blends gracefully into the top of the fuselage.

At the back of the passenger area is a wraparound, rear-view screen and there are three good-sized windows on each side of the cabin. So with the big windscreen there is plenty of light inside. On the left-hand side there is a large door giving access to the front pair of seats. On the other side of the fuselage a quite massive pair of doors opens to reveal a hole of astonishing proportions in a small aircraft. Through it, the four passengers may embark or disembark with relative ease. The seats may easily be removed to make the Stationair 6 an attractive proposition for parachute training.

Up in the sharp end of the ship, hiding behind man-sized cowlings, is a muscle-bound 310-hp engine, its exhaust emerging from under the front of the nose in an effort to keep the noise well away from the cabin. Thrust is provided by a three-blade propeller, 80 in (203 cm) in diameter. The mainwheels are attached to the usual spring steel legs that are a favourite with this manufacturer (at least they need little or no maintenance) and the nosewheel is supported by a telescopic strut. No attempt is made to provide fairings, so the torque links are allowed to dangle in the breeze – which is a pity because otherwise the Stationair 6 is quite a handsome hunk of technology on the ground and in the air.

Cabin and flight deck

The long cabin measures 44 in (112 cm) from wall to wall, giving adequate elbow room for pilot and passengers alike. 180 lb (82 kg) of baggage may be stowed in the area provided behind the rear pair of seats. Like so many of these American singles the instrument panel is set too high and the optional seat with height adjustment is essential for pilots of average height. The flight deck part of the ship follows standard Cessna practice with management of the electric services on a strip below the left-hand panel. Most of the circuit breakers are alongside the switches – a very convenient arrangement. Provision for map/airways manual stowage could be a lot better. There are no annunciator lights.

A deep panel in front of the captain's seat carries the flight instruments and, to their right, the various radionav readouts. A central panel – large enough for a station box, two NAV/COMs, ADF, DME and a transponder – also provides room for an autopilot control box should this be required. Below the avionics stack are three plungers: throttle, pitch and mixture. I have never cared for these plunger-type controls. Levers are far better to handle and, in my view, easier to use, although to some extent the Vernier adjustment (you twist the knobs for small alterations to manifold pressure, rpm, and so on) redresses the balance a little.

Extending down to the floor is a small console. It contains rudder and elevator trim wheels along with their position indicator, the fuel selector, an engine cooling flap lever and a microphone.

The car-type parking-brake handle pulls out from under the left-hand instrument panel. Engine instruments, which are located on the right-hand panel, have been angled towards the pilot – an excellent arrangement which is intended to reduce parallax error and assist in the accurate management of power. The flap switch and its position indicator reside low down to the right of centre (not the most convenient place for the pilot) and nearby are the effective cabin hot/cold air controls.

The standard of interior trim is professional, functional and tidy rather than posh. But this makes it ideal for bush operations in remote areas where the name of the game is 'fly five passengers in and out of an impossible dust strip' – something the Turbo Stationair 6 can take in its stride, even on a hot day.

In the air

Engine starting is standard for a fuel-injected unit and the motor grumbles into life before settling into a deep, contented growl at idling speed. Any Cessna 152 pilot could step into big brother and taxi away without difficulty. On the day of my flight we had full fuel but only two of us on board so take-off was at low weight. I was therefore advised to rotate at only 40 kt. We leapt off the ground shortly afterwards.

Noise level was moderate if not the quietest I have known. 2,500 rpm and 30 in manifold pressure produced an initial climb rate of 1,250 ft/min at 86 kt. Cessna claim 1,010 ft/min for a fully loaded aircraft. One of the talents

Cessna Turbo Stationair 6, big brother of the Skyhawk.

OPPOSITE: *Cessna's Fowler flaps, a masterpiece of engineering that avoids the use of drag-producing external tracking.*

of turbocharged engines is their ability to maintain power at altitude. I timed from 2,000 to 10,000 ft in 6 minutes 26 seconds, representing an average climb rate of 1,244 ft/min. This it could have maintained up to 12,000 ft, which is about the maximum altitude at which full climbing power can be developed.

On the day of the flight 75 per cent power at 10,000 ft (2,400 rpm/29 in manifold pressure) gave an IAS of 135 kt and that trued out as a snappy 156 kt. Cessna recommend an 80 per cent power cruise, which at that level should give 152 kt when the aircraft is at maximum

weight. 65 per cent power will return a still useful 144 kt but even at the higher power setting the aircraft is relatively quiet. If you climb to 20,000 ft, cruising speeds of up to 167 kt are attainable. Now that is pretty rapid for a six-seater with permanently welded-down wheels, but you would have to wear oxygen masks at that flight level and few passengers take kindly to being reminded of old-fashioned dentistry. Incidentally, the Stationair 6 is the same weight as the now-discontinued Piper Twin Comanche, so in providing similar cruising speeds on 20 less hp (without the benefit of disappearing wheels) it must

be regarded as an efficient airframe.

The follow-up flap control is gated at 10, 20, and 40 degrees. Flaps-up we stalled at 46 kt, and the various flap settings produced stalls at 44, 38 and 35 kt respectively. Like most of these types of aircraft there is a large position error at low airspeeds. With a little power you can fly with the ASI finger at the bottom of the scale.

In terms of handling the Stationair feels like a large aircraft. The rudder and elevators are heavy, the ailerons less so and decidedly more effective than those Micky Mouse affairs in the smaller Cessnas. Lateral stability is not very pronounced but only one cycle was taken to correct a 15-kt speed displacement after I had released the wheel. The trimmers are pleasant and progressive in action.

Visibility in the air is good in most directions except while turning, when the down-going wing tends to obstruct the view. However, the Stationair is somewhat larger than the 172/182 models and Cessna have been able to position the pilots so that if they lean forward slightly it is possible to look ahead of the wing leading edge. The aircraft is good for instrument flying; it ploughs along at a determined trot like a big car on a fast road.

Initial approach at 80 kt, followed by 70 kt over the threshold works nicely and like all Cessnas the landing is simple with no last-minute party tricks to master. When the need arises, low-speed approaches using the correct technique can result in some remarkably short arrivals. I liked the Turbo Stationair 6 in the air. It is a steady weight lifter and a pleasure to fly.

Capabilities

Maximum ramp weight is 3,616 lb (1,640 kg) and a well-equipped Stationair would have an empty weight of 2,059 lb (934 kg), leaving a useful load of 1,557 lb (706 kg). Full tanks would leave 1,027 lb (466 kg), which is enough for six 170-lb (77-kg) adults without luggage. It can then fly a with-reserve range of 655 nm at 80 per cent power (10,000 ft). Alternatively, if you adopt the low-power economy setting, 805 nm can be flown at reduced speed. I am confining these figures to flight levels where oxygen need not be used.

A full cabin (six 170-lb (77-kg) people plus 180 lb (82 kg) of baggage) would entail reducing fuel by 24 Imp/29 US gal (110 litres), when the range becomes a still respectable 440 nm.

If you remove the five passenger seats there is an additional 102 lb (46 kg) of useful load, and operated as a light freighter the aircraft would carry about 1,500 (680 kg) over a distance of 440 nm. Obviously the large double doors are ideal for loading bulky packages. A fully loaded aircraft will reach a height of 50 ft in a distance of 1,640 ft from the start of the take-off run when there is no wind; so under normal conditions you could operate a Turbo Stationair 6 out of 600-yard airstrips and the turbocharger makes it ideal for 'hot and high' conditions. Having such a long cabin and those big double doors, the aircraft makes a capable air ambulance.

Verdict

The Cessna Turbo Stationair 6 is a great and versatile work-horse. It is the kind of vehicle you would welcome in those parts of the world where a 10-mile journey by mule or camel might take all day.

With simplicity the keynote of its airframe, the aircraft is ideal for operations far removed from product support. Furthermore, its docile handling and hot/high capabilities make it a natural for such areas as Central Africa, Australia, South America and the like.

The Turbo Stationair 6 is really the odd man out in this chapter since it is not the kind of aircraft most private owners would consider buying. It is, rather, an 'all-can-do' light single, capable of making money in the right hands when applied to proper tasks.

Facts and figures

Dimensions

Wing span	35 ft, 10 in
Wing area	174 sq ft
Length	28 ft, 3 in
Height	9 ft, 3 in

Weights & loadings

Max ramp	3,616 lb (1,640 kg)
Max take-off	3,600 lb (1,633 kg)
Equipped empty	2,059 lb (934 kg)
Useful load	1,557 lb (706 kg)
Seating capacity	6
Max baggage	180 lb (82 kg)
Max fuel	73 Imp/88 US gal (333 litres)
Wing loading	20.7 lb/sq ft
Power loading	11.6 lb/hp

Performance

Max speed at 17,000 ft	174 kt
80% power cruise at 10,000 ft	152 kt
80% power cruise at 20,000 ft	167 kt
Range at 80% power (10,000 ft)	655 nm
Max range at reduced power	805 nm
Rate of climb	1,010 ft/min
Take-off distance over 50 ft	1,640 ft
Landing distance over 50 ft	1,395 ft
Service ceiling	27,000 ft

Engine

Continental TSIO-520-M, turbocharged, fuel-injected producing 310 hp at 2,700 rpm, driving a three-blade, 80-in (203-cm) diameter propeller.

TBO

1,500 hours.

Simple, capable and rugged – a station-waggon of the air.

HOW DO THEY RATE?

The following assessments compare aircraft of similar class.
For example, four stars (Above average) means when rated
against other designs of the same type.

★★★★★ Exceptional
★★★★ Above average
★★★ Average
★★ Below average
★ Unacceptable

Aircraft type	Appearance	Engineering	Comfort	Noise level	Visibility	Handling	Payload/Range	Field Performance	Cruising Speed	Economy	Value for Money
Cessna 172 Skyhawk	★★★★	★★★	★★★	★★★★	★★	★★	★★	★★★	★★★	★★★	★★★
Piper PA-28-161 Warrior II	★★★	★★★	★★★★	★★★★	★★★	★★★	★★★	★★★	★★★	★★★	★★★
Robin DR400/Major 80	★★★★	★★★★	★★★	★★★	★★★★★	★★★★	★★★★★	★★★	★★★★	★★★★	★★★★
Robin R3140	★★★★	★★★★	★★★★	★★★	★★★★★	★★★★	★★★★	★★★★	★★★★	★★★★	★★★★
Aerospatiale TB10 Tobago	★★★★	★★★★	★★★★★	★★★★	★★★★	★★★★	★★★★	★★★	★★★	★★★	★★★★
Cessna Turbo Stationair 6	★★★	★★★★	★★★★	★★★	★★★	★★★	★★★★	★★★	★★★★	★★★★	★★★

4. SINGLE-ENGINE RETRACTABLE TOURERS

The reason for fitting vanishing legs will be well known to pilots and aeronautical engineers; however, for the benefit of those recently afflicted with the aviation bug (an incurable illness if ever there was) perhaps a few words on the subject are in order. For speeds of up to 140–150 kt there is usually little to be gained by fitting a retractable undercarriage provided care is taken to enclose the wheels and all legs in well-designed fairings. Above that speed drag is such that real benefits can follow the addition of retractable gear; although you must pay a penalty in terms of increased weight for the electric or hydraulic system that raises and lowers the wheels. There is also the cost of fitting this extra equipment in the first place and, of course, its subsequent maintenance.

What do we get in real terms from a retractable undercarriage? In round figures, Wonderplane X with a wheels-up cruising speed of 168 kt at 75 per cent power may only be able to manage 150 kt with them down. Now 15 minutes saved on a 350-mile (563-km) journey may not exactly make the headlines but it does represent a 10.7 per cent saving in fuel, and over a year that could be very worthwhile. But the advantages, which become more attractive as aircraft speed increases, do not stop here. There are tangible improvements in rate of climb. This is particularly important in the case of light twins (and some not-so light ones I can think of) where wheels-up or wheels-down is the difference between climbing to safety when an engine fails after take-off and a gentle but persistent descent.

By the time you aspire to flying a single-engine retractable aircraft with its constant-speed propeller, advanced avionics, an autopilot and perhaps even pressurization and de-icing gear, you are involving yourself in equipment that is as complex as that used by the airlines of not so many years ago. Naturally the advantages of a retractable undercarriage bring with them added responsibilities and a need for more professionalism on the part of the pilot – but that is another story.

What is expected of a single-engine retractable? Having regard to the additional complexity and cost of providing vanishing wheels, I would say all the requirements listed at the beginning of the previous chapter – only more so.

Finally, there is the question of whether or not you should spend so much money on an advanced single – because for not a lot more than the price of the more expensive singles you could buy yourself one of the smaller light twins. I shall enlarge on this topic in the next chapter but much depends on the type of routes most likely to be flown by the owner/operator. While engine reliability has certainly improved over the years, engine failures persist on the aviation scene, albeit at a reduced rate. And when the one and only donkey gives up trying, say, over water and miles away from land – ! In other words, unless you live in parts of the world where most flying is over hospitable territory (that is, away from oceans, mountains and jungles) the single-engine tourer is not an ideal business aircraft, particularly when icing risks are high. Few light singles have full ice protection and without it you can be grounded at those times of the year when the icing index is high. In my opinion the high-performance single should be regarded as a luxury touring aircraft.

Disappearing legs. Mooney M20J 201.

Mooney M20J 201

The fast individualist

Background

Early models of the Mooney four-seat tourers, which first flew in 1954, had wooden wings of questionable structural integrity and a nauseating mechanically-operated under-carriage. There was a long lever in the cockpit which had a locking device and required twisting before it could be pulled to retract the wheels.

Fortunately these early eccentricities have gone but a few more remain and these will be described in the course of this report. I use the word 'eccentric' with the qualifica-tion that 'unusual' features can often be good – there is nothing wrong with being nonconformist, provided there is good reason for being out of step.

These days Mooney, of Kerrville, Texas, offer two vari-ants of their M20 design. There is the normally aspirated 201 model and the turbocharged 231. These numbers, by the way, relate to the maximum speed of the aircraft in miles per hour. The subject of this airtest is the Mooney 201.

Engineering and design features

The metal wing has a relatively high-aspect ratio of 7.338 and flush rivets are used from the leading edge to point of maximum camber. Two integral tanks provide a total usable fuel capacity of 53 Imp/64 US gal (242 litres). Here is where we start on the eccentricities. Although Mooney make a big sales point of efficiency (and their aircraft are certainly efficient) there are no wing-tips; they might just as well have taken a saw and cut off the ends of the mainplane – ailerons and all. I find this very strange, considering the trouble they have taken to seal gaps between the control surface and their adjacent fixed struc-tures. That omission must cost a few knots in cruising speed. (I see they fit wing-tips to their 231 model.) Slotted flaps may be set to the following positions: UP, TAKE-OFF (15 degrees) and LANDING (33 degrees).

There is a tailplane with separate elevators and a fin, the leading edge of which sits more or less bolt upright, while the rudder sweeps forward. It looks as though it has been put on back to front. Now we come to another eccentricity. For the purpose of longitudinal trim the entire rear of the aircraft – tailplane, fin and the last few feet of the fuselage – is arranged to hinge up and down. As the tailplane angle of incidence alters, so the fin and rudder lean back and forth when viewed from the side.

The cabin area is constructed of welded steel tubes covered in light alloy panels. From behind the rear seats conventional stressed skin is employed. There is a well-designed engine cowling surrounding the 200-hp Lyco-ming engine and, to minimize drag, the landing light is installed behind a transparent panel which follows the contours of the nose. The fuselage is of good aerodynamic shape but has shallow cabin windows which start high up the fuselage walls and leave a deep roof. Yet another drag saver is an electronic OAT gauge sensor which replaces the usual temperature probe.

The aircraft sits low on its rubber-in-compression under-carriage, so no steps are required to assist entry onto the

right-hand wing, where a walkway is provided adjacent to the single entrance door. Quality of engineering is above average but in my opinion the Mooney is spoiled by its mean little windows and back-to-front tail.

Cabin and flight deck

You enter the back of the cabin by tilting the front seats. A no-slam door is shut by gently closing and allowing the lever to pull it in hard against its weather seals. At 43 in (109 cm) wide the cabin is not unduly narrow but the roof is low and the small high-set side windows make the interior look smaller than it actually is. Personally I found the Mooney 201 claustrophobic, particularly when all four seats are filled, which is a pity because the interior is beautifully appointed with good-quality trim and a well-designed control layout.

Six square instruments are lined up along the top of the left-hand panel, two of them reading fuel quantity in pounds. To their right are similar readouts for fuel pressure, oil pressure, oil temperature and cylinder head temperature. Directly below are the flight instruments, the example tested having a rather fussy true airspeed indicator, smothered in figures because our American cousins are fighting a rearguard action in support of miles per hour. Personally I sympathize with them. No one, in my view, has ever put up a convincing argument in favour of knots, but the decision to use them for measuring speed has been agreed internationally. If it was not for the long-established use of knots by the international marine fraternity we might have been stuck with the kilometres per hour (the French are fighting a rearguard action on behalf of those, by the way), so we should all be thankful for what we have got. For no extra charge, you can buy a Mooney with a real ASI calibrated in the international standard units.

Radionav readouts and the avionics take up the centre

Mooney M20J 201. Shallow, high-set windows, short undercarriage legs and a back-to-front fin and rudder.

and, if required, extend over much of the right-hand panel. Far right is a small area, conveniently angled towards the pilot, which displays the fuel flow/manifold pressure gauge, the rpm indicator and all the circuit breakers neatly laid out and labelled in seven rows.

A console extends down to the floor from the centre of the panel. At the top are three Vernier plungers for the throttle, propeller and mixture; and pull-out knobs for alternate air, the parking brake and the engine cooling flaps. Lower down is the three-position flap lever and its indicator, which is side by side with the elevator-trim position indicator. Below is the autopilot control panel and at the bottom of the stack are three knobs for cabin air and temperature, which work effectively for all occupants. The undercarriage is handled on a wheel-shaped switch located in the centre of the panel just under the glareshield, and the electrics are managed on a row of rocker switches along the bottom of the captain's panel. Stowage for maps and manuals is average and that means not good enough.

Other than the split engine instruments (some far right, others top left) I thought this was an excellent flight deck. However, now we come to yet another oddity. I am of at least average height, perhaps a shade taller, yet I could only reach the rudder pedals by sliding forward the seat until the control wheel and instruments were uncomfortably close. Mooneys are built in Texas where many of the locals are so tall they use oxygen when standing up. But the rest of us have to live too and I suggest that if Mooney feel unable to reposition their rudder pedals they should make them adjustable for leg reach.

A more attractive feature is the baggage door which is located high up behind the cabin on the right-hand side. It is large enough to be used as an emergency exit. Stowage space for up to 120 lb (54 kg) of luggage is provided, not exactly mind-blowing for a 200-hp tourer but adequate for three people going in style or four who are prepared to do their own laundry once they arrive.

In the air

The 200-hp Lycoming engine is fuel injected and is fired up accordingly. While taxiing, the aircraft goes where you expect it to, the brakes are powerful and the nosewheel steering is reasonably light. Visibility ahead is satisfactory, but somewhat restricted to the right by rather thick door pillars. The aircraft I was testing had full fuel and we were within 200 lb (91 kg) of the 2,740-lb (1,243-kg) maximum take-off weight.

I timed twelve seconds from start of the roll to lift-off. There was plenty of noise but not to the point where conversation had to cease. Initial rate of climb was a lusty 1,300 ft/min, which is about 270 ft/min better than claimed for a fully loaded example, although I don't believe another 200 lb (91 kg) in the cabin would have made all that difference (perhaps 100 ft/min at the most).

At 4,000 ft, 75 per cent power (2,600 rpm/24 in manifold pressure) presented us with a TAS of 162 kt but the place to be on that power setting is at 8,000 ft where there is a

When the elevator trim wheel is adjusted, the tailplane, fin and rudder, and rear fuselage move accordingly.

OPPOSITE: The interior is nicely trimmed and there is a reasonable amount of room despite the somewhat confined atmosphere.

spirited 169-kt cruise at maximum weight. Now that has got to be good news for a 200-hp tourer. I have known better noise levels – but I have also known worse.

Mooney have reversed the ideal input proportions by making the ailerons heaviest and the rudder lightest – back to eccentricities again. Despite this, the 201 is fun to fly and if handling is unusual it nevertheless feels like an aircraft – not a Chev-on-wings. Even the electric trimmer works at a sensible rate. So many light-aircraft trimmers creep around as though they are on their last legs (as opposed to those on large jet transports, which threaten to have your leg off like an electric saw). The aircraft also does well in terms of stability, which is good in all three axes. Although they should lighten the ailerons and heavy-up the rudder I rate the Mooney as above average in handling and stability.

Wheels-and flaps-up the stall came at 62 kt with a gentle

left-wing drop. With flap the figures were: 15 degrees, 50 kt; and 33 degrees, 45 kt, wings level. Visibility in the air is a little restricted, particularly to the right.

Short finals are flown at 70 kt. You must remember that the Mooney sits low on the ground, otherwise there are no problems during the landing. The elevators remain powerful to the last and you should bear that in mind. Your first landing might not be perfect, but the next will be a greaser.

Capabilities
On the same power as a Piper Arrow IV the Mooney 201 is 28 kt faster at 75 per cent power, 200 ft/min better in rate of climb and it will burn 23 per cent less fuel on the same journey. A lavishly equipped version would have an empty weight of 1,785 lb (810 kg) and provide a useful load of 955 lb (433 kg). With full tanks there is a payload

of 573 lb (260 kg) – say three 170-lb (77-kg) adults and 63 lb (29 kg) of baggage or two 160-lb (73-kg) men and two 128-lb (58-kg) women. Allowing for the usual reserves, full tanks will fly a Mooney 201 for 700 nm at 169–171 kt according to weight. Alternatively, you could cover more than 850 nm by adopting a 55 per cent setting for a still respectable 150 kt. A full cabin – four 170-lb (77-kg) occupants plus the 120 lb (54 kg) maximum baggage – would allow 155 lb (70 kg) for fuel – 21 Imp/26 US gal (98 litres) – when the range becomes 280–330 nm according to power setting and cruising level.

Mooney rightly plug the aircraft's efficiency when promoting the 201. So they should. It is 16 kt faster at 75 per cent power than the Cessna Skylane RG, which enjoys the advantage of another 35 hp. It can even match the 285-hp Beech Bonanza F33A which costs almost twice as much to buy.

The distinctive Mooney has a very efficient airframe but the cabin would be improved if larger windows could be provided.

Verdict

With so many odd features built into it you cannot be indifferent to the Mooney 201. Either you love it or you hate it. Mooneys have their passionate supporters as well as their detractors. Frankly, much as I admire its efficiency – and a lift/drag ratio of eleven has much to do with this – the rather chummy atmosphere of the cabin, the built-in claustrophobia (those high-set little windows) and the misplaced rudder pedals switch me off. But having been more than a little rude about the nonconformist nature of the 201 I must admit to admiring its above-average engineering, high-quality interior, excellent air miles per gallon and likeable behaviour on the ground and in the air. Another plus mark is that most aviation writers are agreed that Mooney tend to understate their performance figures.

I look forward to the day when they put some real windows in it and fit adjustable rudder pedals. Then (and only then) will I forgive them for their rear fuselage that bends up and down when you work the elevator trim.

Facts and figures

Dimensions

Wing span	35 ft
Wing area	167 sq ft
Length	24 ft, 8 in
Height	8 ft, 4 in

Weights & loadings

Max take-off	2,740 lb (1,243 kg)
Equipped empty	1,785 lb (810 kg)
Useful load	995 lb (451 kg)
Seating capacity	4
Max baggage	120 lb (54 kg)
Max fuel	53 Imp/64 US gal (242 litres)
Wing loading	16.4 lb/sq ft
Power loading	13.7 lb/hp

Performance

Max speed at sea level	175 kt
75% power cruise at 8,000 ft	169 kt
65% power cruise at 10,000 ft	160 kt
Range at 75% power (8,000 ft)	700 nm
Range at 65% power (10,000 ft)	756 nm
Rate of climb	1,030 ft/min
Take-off distance over 50 ft	1,550 ft
Landing distance over 50 ft	1,980 ft
Service ceiling	18,800 ft

Engine

Lycoming IO-360-A3B6D developing 200 hp, driving a McCauley 74-in (188-cm) diameter, two-blade, constant-speed propeller.

TBO

1,800 hours.

Aerospatiale TB20 Trinidad

The flying Jaguar: space, pace and grace

Background

I dealt in the last chapter with the TB10 Tobago, a product of the Aerospatiale light-plane division in France. This report is about a higher-powered retractable version of the same airframe.

When the TB9/TB10 series was first contemplated it was always envisaged that more complex versions would be developed, leading eventually to a light twin. The first of these developments is the Trinidad. In essence the airframe is, rivet for rivet, the same as that described in the Tobago airtest, with a few exceptions that will be dealt with in this airtest.

The Trinidad was introduced in 1981, not long after the demise of the Rockwell Commander 112 and 114 single-engine retractables, and about nine years after production of Piper's excellent Comanche 260 had ceased. So the opposition as it now stands is either less powerful – 200-hp Piper Arrow, 235-hp Cessna Skylane RG, 200-hp Mooney (see previous test), or larger and more powerful, 285-hp Beech Bonanza, 310-hp Cessna Centurion, and so on (see later in this chapter). With its 250-hp engine, the Trinidad is more or less in a class of its own.

Engineering and design features

The Trinidad wing is the same as the previously described Tobago's. It is perhaps unique among light planes in having a main spar machined from solid metal on a computer-controlled milling machine. The technique has long been a part of large transport and military aircraft production but I believe the Aerospatiale light planes to be the first with such an advanced mainspar. The advantages of a homogenous structure, free of joins or rivets, are obvious.

Pick-up points for the retractable undercarriage are incorporated in the spar, whether or not it is destined for a Trinidad or its fixed-gear brother. Retraction and lowering is hydraulic, pressure being provided by an electro-hydraulic power pack. The wheels are locked up by hydraulic pressure which is maintained by an automatic valve. Doors close behind the undercarriage as it retracts. An overcentre stay on each leg locks the wheels down. The two main legs have microswitches which prevent the pilot from selecting WHEELS UP while on the ground – an act likely to promote him to idiot of the week. There is a simple emergency lowering system which breaks the hydraulic pressure, allowing the wheels to drop down and lock by gravity assisted by springs.

Two integral fuel tanks with a total usable capacity of 72 Imp/86 US gal (326 litres) are built into the wings. The

tail surfaces are similar to those of the lower-powered versions, except for a rudder trim tab which has been added to, rather than set into, the trailing edge. In every other respect the airframe is as described in the Tobago airtest in Chapter 3.

Cabin and flight deck

Along the top of the captain's flight panel is the annunciator, a strip of coloured warning lights which, from left to right tell of the following: alternator OFF, fuel OFF, parking brake ON, oil pressure warning, a press-to-test button which illuminates all lights for a bulb check, a day/night dimmer switch, fuel pump ON, pilot heat ON, taxi lights ON, landing lights ON. The undercarriage lights

Aerospatiale TB20 Trinidad, top-of-the-line light single from this French manufacturer.

(three greens for locked down, a single red when they are not) are positioned to the right and below the captain's control yoke. The emergency lowering knob is below the left-hand fresh-air vent. There are those very neat thermal overload switches for the electric services (press the green button for ON and the red one for OFF), which are identified by so-called international signs, reminiscent of ancient Egyptian hieroglyphics. The switches are conveniently located in front of the power controls on the central pedestal which runs back between the pilots. It carries the throttle, pitch and mixture levers, the rather stiff elevator trim wheel and its indicator, and a cigar lighter (pilots are assumed wealthy enough to smoke them – so it is never referred to as a cigarette lighter in the flight manual). At

the rear of the console is the rudder trim knob with a scale marked for quick selection of the best position for TAKE-OFF, CLIMB and CRUISE. Everything else is pretty well the same as in the Tobago and I can only repeat that this is a very pleasant 'office' and quite unlike any other light aircraft.

In the air

The 250-hp Lycoming is a fuel-injected unit, and, having primed it with the booster pump, the mixture control is returned to the CUT-OFF position prior to operating the

starter key. As the engine fires, the mixture is moved forward to the fully RICH position.

Taxiing is like any modern light plane, no more demanding than steering your car. Take-off flap (10 degrees) is selected, the rudder trim is moved to the TAKE-OFF position (three divisions right) and the elevator trimmer is set to neutral. There is little tendency to swing as power is added and at 67 kt the control yoke is brought back, the aircraft lifts off and an initial climb is established at 72 kt. At a safe height the undercarriage is retracted and the flaps are raised at 300 ft. Recommended climbing speed is 92

kt and the VSI gave a reading of about 1,300 ft/min for an almost fully loaded aircraft, which is lively by any standards. At 4,000 ft, 2,300 rpm/24.4 in manifold pressure, representing the 75 per cent power setting at that altitude, gave a TAS of 158 kt; and 65 per cent power, entailing a manifold pressure of 21.6 in, flew us at 147 kt. Of course, like most normally aspirated aircraft the Trinidad is at its best while cruising in the 8,000–12,000-ft band. For example, 75 per cent at 8,000 ft (the maximum altitude where this is attainable) will fly a maximum-weight TB20 at a cracking 164 kt and if you climb to 12,000 ft the 65

Circuit breakers are on a stepped panel by the captain's left knee, engine temperatures and pressures are shown on the vertical readouts at the top of the radio stack.

As a fast tourer the Trinidad has a lot to offer. For more serious use, ice protection would be needed.

per cent setting is good for 158 kt. I am quoting figures for 'best power'. But if you are prepared to trade a few knots speed in return for economy the 65 per cent setting will fly you and your guests at 155 kt for a fuel burn of only 72.6 lb (32.9 kg) an hour. In motoring terms we are talking of flying up to five people in a big cabin at almost 180 mph (three miles a minute) while returning 14.79 miles per US gal or 17.72 miles per Imp gal. No family car can cruise at that kind of speed and the two-seat, hairy GT jobs capable of in excess of 150 mph would probably not even achieve a miles per gallon figure that ran into

double numbers.

The Trinidad is 415 lb (188 kg) heavier than the Tobago. It is also more than 30 kt faster in the cruise, consequently it has the feel of a light twin. I can best describe the ride as rigid and on-the-rails. Even without an autopilot it should prove relaxing to fly on relatively long journeys with its excellent visibility through those big windows, the above-average stability and spacious cabin.

One area that should be improved is cabin noise. The lower-powered Tampico and Tobago are quieter than most of the light singles on the market but the 250-hp

Lycoming tends to be rather vocal and those extra knots create a fair amount of airframe noise. There is nothing new about these problems but they cannot be ignored. I would like to see an improved exhaust system for the Trinidad, better cabin insulation in the roof and walls, and double windows. (Piper did this on the old Twin Comanche and it worked wonders.) I dwell on the subject of noise because the Trinidad is so outstanding in most respects that it deserves a quieter cabin. At present I would describe the noise level at 75 per cent power as average. But that is not good enough.

The clean stall came at 65 kt; maximum flap (40 degrees) took 10 kt off that figure and provoked a slight tendency for the right wing to drop, which did not develop. No doubt because of the higher speeds, control loads are appreciably heavier than in the Tobago. This was particularly noticeable with the ailerons, although the example tested seemed to be suffering from excessive control-run friction. It was later discovered that this had been caused by a maladjusted autopilot servo. The stabilator, rudder and ailerons are moved with push-pull rods.

Initial approach may comfortably be flown at 85 kt aiming for 70 kt over the fence when at or near maximum weight. After the round-out the throttle is closed and you must be prepared to start holding off immediately. Otherwise the hardware will sink onto the runway in the level altitude and arrive all wheels together with a tendency to skip. Remember that and you will find the Trinidad as easy to land as any simple light plane.

Capabilities

Maximum ramp weight is 2,950 lb (1,338 kg) and an equipped example would turn the scales empty at around 1,750 lb (794 kg), leaving a useful load of 1,200 lb (544 kg). Maximum usable fuel would account for 518 lb (235 kg), allowing a payload of 682 lb (285 kg) – enough for four 170-lb (77-kg) adults without baggage or two men at 160 lb (73 kg), two women at 126 lb (57 kg) and the not very generous 110-lb (50-kg) maximum baggage allowance. So loaded, our Trinidad would fly a with-reserve 800 nm at 164 kt, 880 nm at 158 kt and a long-range ferry at 55 per cent power of more than 1,100 nm. So for practical purposes this is a full-tanks/full-cabin aircraft unless you want to carry three passengers on the back seat. Then you would have to do without baggage and reduce the fuel uplift by 30–60 lb (14–27 kg) according to the weight of the extra person. I would regard the Trinidad's payload/range capabilities as above average for its class.

In terms of field performance the TB20 does well for itself. Compared with the much lighter and lower-powered Piper Archer (180 hp) its field requirements are only 100 ft more. And it demands 595 ft less take-off distance than the Beech Bonanza A36 (described later in this chapter). You can operate a Trinidad comfortably in and out of a 600-nm strip unless your local airfield is 'hot and high'.

Verdict

The cynics may say: 'So what's special about the Trinidad? It's no faster than Piper's single-engine Comanche which first appeared in the late 1950s.' I would answer that the Trinidad's cabin is light years ahead of the Comanche in terms of space, visibility and design, also the aircraft handles better and is more forgiving.

I would like to see a bigger door to the baggage compartment, a larger baggage allowance, a nicer handling elevator trimmer and a reduction in noise level. But it is early days for the remarkable Trinidad. With time and production experience even indifferent aircraft have been known to become acceptable. And with very little effort, the above-average Trinidad could develop into one of those outstanding light planes that grace the aviation scene from time to time

Facts and figures

Dimensions

Wing span	32 ft
Wing area	128 sq ft
Length	25 ft
Height	10 ft, 6 in

Weights & loadings

Max ramp	2,950 lb (1,338 kg)
Max take-off	2,937 lb (1,332 kg)
Equipped empty	1,750 lb (794 kg)
Useful load	1,200 lb (544 kg)
Seating capacity	4/5
Max baggage	110 lb (50 kg)
Max fuel	72 Imp/86 US gal (326 litres)
Wing loading	22.94 lb/sq ft
Power loading	11.75 lb/hp

Performance

Max speed at sea level	169 kt
75% power cruise	164 kt
65% power cruise	158 kt
Range at 75% power	800 nm
Range at 65% power	880 nm
Rate of climb	1,260 ft/min
Take-off distance over 50 ft	1,571 ft
Landing distance over 50 ft	1,740 ft
Service ceiling	20,000 ft

Engine

Lycoming 10-540-C4D50 developing 250 hp at 2,575 rpm, driving a Hartzell two-blade, constant-speed propeller, 80 in (203 cm) in diameter.

TBO

2,000 hours.

Siai Marchetti SF260

The poor man's Spitfire

Background

The Italian aircraft firm, Savoia Marchetti, was formed in 1915. It built flying boats, civil airliners and the very potent SM79 bomber which appeared in 1936. This had three engines and some of the fighters of the day were hard pressed to catch it. There were mergers, a change of name to Siai Marchetti and all manner of activities, which included helicopter development in conjunction with Augusta. In 1981 there was another shuffle and Industria Aeronautica Meridionale was formed under a single management with Siai Marchetti in the lead.

Meanwhile, back in the early 1960s a brilliant Italian designer named Sterlio Frati came up with an exciting high-performance tourer, the F250, and a prototype was built by a small outfit calling itself Aviamilano. It made its first flight on 15 July, 1964 and, with minor changes, Siai Marchetti built it under licence as the SF260. FAA approval was obtained on 1 April, 1966 and at that point Siai Marchetti took over the design.

There have been private-owner, civil and military trainer and light ground-attack versions of the SF260, and by 1982 over 800 had been sold, despite the fact that it is one of the most expensive light singles on the market.

Engineering and design features

The SF260 is very much in the tradition of those mouth-watering Italian GT cars: small, sexy and very fast. It is one of the most beautifully proportioned single-engine designs imaginable, with a small tapered wing of only 108.7 sq ft area. The manufacturers describe it as a four-seater. So it is in the Italian sports car sense; but you would only ask your two worst enemies to occupy the back seats, which are limited to a total maximum load of 250 lb (113 kg). So the aircraft is really a three-adult tourer or a '2 plus 2'.

Two small wing tanks have a combined capacity of only 21 Imp/25 US gal (95 litres), but there are tip tanks which bring total fuel to 51 Imp/61 US gal (231 litres) usable. Electrically operated slotted flaps are fitted which proved to be more effective than usual for this type of high-lift device.

A long dorsal fin starts from the back of the canopy and

Siai Marchetti SF260, a private fighter of a light plane, with thin, high-speed wings. Most of the fuel is carried in tip tanks.

meets up with the swept fin and rudder. There is a tailplane and separate elevators. The elegant fuselage is crowned by a large, one-piece canopy which slides back to reveal the entire cabin area. There are three guide tracks: one on each side and another let into the top of the dorsal fin. A 260-hp carburettor-type Lycoming engine resides within the shapely nose. The entire masterpiece sits on an electrically retracted undercarriage. Copious use is made of flush riveting and this +6g −3g airframe is the nearest thing to a private Spitfire you are ever likely to see. It is as smooth as the proverbial baby's bottom. Standards of engineering are very high throughout, but that is hardly surprising because, over the years, Siai Marchetti have beefed up the structure to please the military who, in some parts of the world, have fitted these aircraft with bombs, guns and rockets.

Cabin and flight deck

Having climbed onto the wing and pulled back the canopy, it is easy to enter the cabin which at 42 in (107 cm) offers the same width as a Beech Bonanza, 2 in (5 cm) less than the Cessna Centurion and 8 in (20 cm) less than the Aerospatiale Trinidad (see airtests in this chapter). The interior is an odd mixture of the plush and the military. Furnishings are plush but the flight panel could have been stolen from a service aircraft. There is a bench seat at the back with space behind it for up to 90 lb (41 kg) of baggage, which is adequate for two people but only just enough for three. Individual seats up front give excellent support and there is a good fresh-air/cabin, heat system, which works well for pilot(s) and passenger(s). Map stowage is minimal.

The black instrument panel is crammed with dials and old-fashioned (but excellent quality) tumbler switches. At top centre of the panel are the wheel-shaped undercarriage switch and its position lights, a key-type ignition/starter switch, battery master and generator switches, and the flap control. Engine instruments are below. Avionics are spread about – and that I do not care for. Some occupy the bottom of the centre panel while others are on the right; and the station box, which allows you to select which radio will be heard through the headset and which transmitter is in use, is above the flight panel on the left. Switches for the electric services are bottom right. There is a parking brake toggle high up on the left-hand panel. A small quadrant protrudes from the centre of the panel and makes a home for the throttle, pitch and mixture levers. Directly below is the fuel selector. Well-designed control sticks are provided.

I do not object to the slightly illogical groupings on the instrument panel in this case – it gives character to this very unusual light plane. But I deplore the two fuel gauges which double for left and right tanks at the flick of a switch. They are about the size of postage stamps and not all that easy to read. Siai Marchetti should devote more space to fuel contents and reduce the size of the incongruous fuel pressure gauge which is almost the size of an

altimeter. Surprisingly for a relatively complex aircraft, there is no annunciator.

In the air

Starting is typical for a carburettor-type engine and a combination of excellent all-round visibility, powerful brakes and very light nosewheel steering makes the SF260 a pleasure to taxi. With a wing loading of over 22 lb/sq ft this is a relatively potent aircraft in light-aircraft terms and flap 20 degrees is needed for the take-off. Being small and quite heavy, it sits firmly on the runway and as full power is applied acceleration is purposeful, straight and true. Lift-off at the recommended 70 kt came after 12 seconds, and 1 minute from start of the roll we hit 1,850 ft with the VSI holding a steady 2,000 ft/min. Noise level throughout the take-off and climb was average – not good, but toler-

able. At 5,500 ft, 75 per cent power produced an IAS of 174 kt (188 kt true), but to get the best cruise performance you should climb to 7,500–10,000 ft. For example, an economical 58 per cent power will fly a fully loaded SF260 at 176 kt for a fuel burn of 9.8 Imp/11.7 US gal (44.3 litres) per hour, while 74 per cent power at 7,500 ft will result in a cracking 192-kt TAS. There is very little engine noise at these speeds but, not surprisingly, you are aware of some airflow hiss around the canopy.

The stall is totally without temperament for so fast an aircraft. Flaps-up the 'g' break came at 66 kt, flap 20 degrees reduced it to 61 kt and with full-flap (50 degrees) it stalled at only 55 kt, with a slight tendency for the left wing to drop.

Visibility in the air through that magnificent, distortion-free canopy, is excellent through 360 degrees. Handling is

The cockpit could be that of a military aircraft built in the 1950s.

superb, with light, precise controls, slightly marred in the example flown by excessive aileron friction which probably required some adjustment. Stability all round is of a high order, with slow, highly damped pitch corrections. Designers at the big GA factories should be made to fly the SF260. They could learn a lot from its exceptional behaviour in the air.

The SF260 is cleared for spinning and aerobatics at a maximum weight of 2,205 lb (1,000 kg) – that means with the rear seat unoccupied and tip tanks empty. Spin entry is slow and deliberate, the nose taking on a steep attitude. During the recovery, spinning stopped after I had applied

full corrective rudder and moved the stick forward just a few inches.

You can enter most aerobatic manoeuvres from the cruise on this remarkable private fighter, typical indicated speeds being 166 kt for the loop, 185 kt for a roll off the top of a loop and 145 kt for a slow roll. Cuban eights and barrel rolls may easily be flown at cruising speed. The SF260 has no inverted flight system so it would be unsuitable for competition aerobatics. But more graceful than the push and grunt of snarling little biplanes are those smooth, sweeping art-in-sky manoeuvres that are sheer poetry to the discerning. And no aircraft is better at these than the SF260.

Since the aircraft flies at double the speed of most trainers you must slow down to a gallop before entering the circuit. At 130 kt, flap 20 degrees may be lowered and soon afterwards the wheels can be extended. A typical base leg speed would be 90 kt followed by short finals with full-flap at only 70 kt. The excellent visibility and precise handling induce a feeling of security on the way in and the landing demands no special piloting skills. The powerful brakes will materially shorten the landing roll.

Capabilities
As a tourer the SF260 has a maximum weight of 2,430 lb (1,102 kg). The lavishly equipped example flown for this airtest had an empty weight of 1,478 lb (670 kg) and a useful load of 952 lb (432 kg). Fill the four tanks and you are left with a payload of 585 lb (265 kg), which is enough for three 170-lb (77-kg) adults and 75 lb (34 kg) of baggage. Or you could take two 170-lb (77-kg) adults and two smaller people on the back seat with a total weight of 250 lb (113 kg). Full tanks will fly a with-reserve 820 nm at 70 per cent power which at 5,000 ft produces a TAS of 186 kt. Even more economical is 58 per cent power, which at 7,500 ft will fly you at 176 kt over a range of 940 nm. Maximum cabin load (340 lb (154 kg) in the front seats, 250 lb (113 kg) in the rear and 90 lb (41 kg) of baggage) means limiting fuel to 272 lb (123 kg), 38 Imp /45 US gal (170 litres), when ranges of 610–700 nm are available according to cruise technique.

Whatever way you look at it, the SF260 is a capable aircraft: superb aerobatics, high cruising speeds and excellent payload/range.

Verdict
The Siai Marchetti is very expensive. You could buy two Aerospatiale Trinidads for the price of one SF260. Quoting another comparison, a Rolls-Royce car costs three times as much as a Jaguar. Now as a Jaguar owner I would question whether a Rolls-Royce is worth three Jags but when I look at the SF260, assessing relative values is not so simple – because here we have a light single that can match or even better twin-engine cruising speeds at much

OPPOSITE: As a fast tourer the SF260 shines; for aerobatics it is sheer poetry.

lower operating costs. It is about 20 kt faster than the Bonanza (see airtest after next). It will even outpace the Pressurized Centurion (see airtest at the end of this chapter). However, like the Beech offering, the latter will carry six people, while the SF260 is really a three-seater or a '2 + 2'. One must acknowledge that without ice protection the SF260, or for that matter any aircraft, is unsuited to all-weather operations. Most light twins may carry de-icing; few singles actually have it fitted. So what is the case for buying the SF260? Well if you want to fly at more than 175 kt for a fuel burn of only 70 lb/hr (32 kg/hr) then this is the only one for you. And if in addition you enjoy the art of aerobatics then, loud and clear, this is the only one. In terms of engineering I cannot fault the SF260, but, be warned, even more difficult than finding the high purchase price of this fabulous aircraft will be dealing with the manufacturers. In my experience Siai Marchetti do not answer letters. Fancy a trip to Italy?

Facts and figures

Dimensions
Wing span	27 ft, 5 in
Wing area	108.7 sq ft
Length	23 ft
Height	7 ft, 6 in

Weights & loadings
Max take-off	2,430 lb (1,102 kg)
Equipped empty	1,478 lb (670 kg)
Useful load	952 lb (432 kg)
Seating capacity	'2 + 2'
Max baggage	90 lb (41 kg)
Max fuel (usable)	51 Imp/61 US gal (231 litres)
Wing loading	22.4 lb/sq ft
Power loading	9.3 lb/hp

Performance
Max speed at sea level	200 kt
74% power cruise at 7,500 ft	192 kt
70% power cruise at 5,000 ft	186 kt
58% power cruise at 7,500 ft	176 kt
Range at 70% power	820 nm
Range at 58% power	940 nm
Rate of climb at sea level	2,000 ft/min
Take-off distance over 50 ft	1,411 ft
Landing distance over 50 ft	2,000 ft

Engine
Lycoming 0-540-E4A5 developing 260 hp at 2,700 rpm, driving a two-blade, 76-in (193-cm) diameter, constant-speed propeller.

TBO
2,000 hours.

Piper PA-32R-301 Saratoga SP

A limousine with wings

Background

Piper are past masters at concocting new aircraft from existing designs. You can imagine the design team, clean newspaper on the table, cigarette ash falling like dandruff and the floor piled up with empty beer cans. Heads are bent in earnest conversation, pocket calculators glow red hot while outside in the real world the bars have long since closed. Suddenly the chief designer announces 'I've got it! We take the wings from the Warrior and add bigger tips to increase the area, use the Lance fuselage and the flaps from the old Cherokee. Add a retractable undercarriage and a 300-hp Lycoming engine and we shall call it the Saratoga SP.' Now there is a lot to be said for toy construction-set engineering where off-the-shelf bits of known reliability are put together and – hey presto! – we have a new wonderplane. The advantages in engineering and production terms are obvious but there is always a risk that the final result will look what it is – a lash-up of parts from other aircraft. In some respects you could level that criticism at the Saratoga SP, but in the main Piper seem to have turned out a successful light aircraft which may fairly be described as a 'limousine with wings'.

Engineering and design features

The wing has a parallel centre portion to which is fitted a pair of tapered outer panels. It is very similar to the Warrior mainplane (see Chapter 3), except that the area has been increased by extending the tips outboard of the ailerons. These, by the way, are proper ones of Frise design. The usual Piper slotted flaps, which in my view could do with generating a little more drag, are mechanically operated with settings gated at UP, 15, 25 and 40 degrees. There are four fuel tanks in the wing and their total usable capacity is 85 Imp/102 US gal (386 litres).

The undercarriage is retracted and lowered by an electro-hydraulic power pack controlled by a wheel-shaped switch on the instrument panel. Retraction and lowering takes about seven seconds but the Saratoga SP has one of those too-clever-by-half automatic gadgets that lowers the wheels irrespective of where you have the selector when the airspeed has dropped below 103 kt, power off. Piper are great ones for this kind of thing. They also have a passion for springs in the control runs; down springs in the elevator (to mask stability weaknesses in pitch), springs between the rudder and the ailerons because (so they believe) the pilot may be incapable of flying a balanced turn. I am very much against this 'planes-for-suckers' approach because suckers should not be flying – and certainly not a 300-hp, seven-seat mini airliner like the

Saratoga. I might be able to show a little more respect for this automatic 'wheels-down' gadget if it had more brains. You can do what you like with the undercarriage control but the wheels will refuse to retract until you are showing at least 81 kt on the clock with full power (and even faster readings at lower power settings). Imagine being faced with a wheels-down ditching soon after take-off. True they provide you with another lever to inhibit the system, but this is adding to the general complication of what is really a simple enough function, that of managing a retractable undercarriage. Perhaps Mr Piper can be persuaded that planes are for pilots and suckers should stick to cycles.

What appears to be a Warrior stabilator and a Warrior fin and rudder make up the tail surfaces. They enjoy the

Piper Saratoga SP, offering twin-engine space with single-engine economy.

advantage of a good lever arm because the Saratoga has a long fuselage. It has to be long because inside are three rows of seats. Four cabin windows are let into each side and up front on the right there is a crew/passenger door. At the rear of the cabin area on the left-hand side is a massive double door that opens to reveal a hole of astonishing proportions. Its 53-in (135-cm) width is of great value when the Saratoga is used as a small freighter. 100 lb (45 kg) of baggage may be stowed behind the rear pair of seats and up front, near the engine firewall, is a separate baggage area with a capacity of 100 lb (45 kg), its access door being let into the right-hand side of the nose. The aircraft is powered by a fuel-injected Lycoming engine delivering a lusty 300 hp to a two-blade propeller (a three-blade one is available as an option).

Viewed on the ground the Saratoga is quite a hunk of aeronautical hardware. But you should remember that, in effect, this is a single-engine Seneca (see Chapter 5), because both aircraft use the same fuselage.

Cabin and flight deck

The example I tested was trimmed in a rather lush red velour material. The chairs had high backs with headrests, there was a hot and cold refreshment console between two of the passenger seats, the four armchairs behind the two for the pilots being arranged 'club' fashion, that is, with those behind the pilots facing rearwards towards the back pair. The cabin is a roomy 49 in (125 cm) wide, and with the overhead spine carrying individual reading lights and fresh-air vents you almost expect an air stewardess to appear with drinks all round. I think it would not be overstating it to describe Piper's deluxe interiors as opulent.

The instrument panel is wide, uncluttered and impressive. In the centre there is room for a double stack of avionics and to the left is the captain's flight panel. Another one could be fitted on the right if required. All engine instruments and circuit breakers are on the lower half of the panel. A small box attached to the left wall carries the master switch as well as those for the electric fuel pump, anti-collision beacon, landing lights and pitot heat.

A quadrant, mounted between the pilot and the right-hand seat holds the throttle, mixture and propeller levers. There is a rudder trim knob of the spring bias type. Below, and near the floor is the fuel selector. They have placed

the elevator trim wheel in a separate quadrant between the front seats along with the manual flap lever which will instantly be recognized by Cherokee pilots (as will the handbrake level with its lock-on button). Next to the flap lever is a small one for overriding that automatic wheel-lowering device. A good annunciator is provided. Although there are map pockets, no provision is made for airways manuals.

The cabin and flight deck areas are very comfortable, nicely planned, spacious and easy on the eyes. What more could you ask?

In the air
Engine starting is as usual for a Lycoming fuel-injected motor. Visibility while taxiing is good – above average by American standards – the brakes are effective and so is the nosewheel steering, although I thought it was a little heavy.

OPPOSITE: *The passenger area is like a mini-airliner.*

From some angles the Saratoga displays its Warrior family connections.

The take-off could not be simpler. Rotate speed varies between 74 and 80 kt according to weight, but for short-field performance flap 25 degrees is lowered and you bring back the control wheel at 58–66 kt, again according to weight. Piper suggest that pilots may prefer to set the cut-out for the automatic wheels-down system before starting the take-off run so that the undercarriage will retract when told to do so by the pilot. I rest my case! I timed from 2,000 ft to 3,000 ft in 47 seconds, representing a healthy 1,276 ft/min rate of climb, which is about 260 ft/min better than claimed for a fully loaded Saratoga. At 4,000 ft, 75 per cent power (2,300 rpm/25.6 in manifold pressure) gave a TAS of 166 kt. This is appreciably above book value but we were about 300 lb (136 kg) below maximum weight at the time. 65 per cent power is obtainable at 2,100 rpm. At this setting the never-noisy Saratoga becomes pleasantly quiet for a 300-hp single, while batting along at a useful 156 kt. Low noise levels are another of the aircraft's plus marks, and in-flight visibility is good by American standards.

I thought the controls were a little firm but the Saratoga is nevertheless a pleasure to fly. No rudder is required for balanced turns and the general handling is honest if not in the top league. The clean stall came at 65 kt and full-flap (40 degrees) showed a worthwhile 11-kt reduction. With flap there is no tendency to drop a wing during the stall. There is powerful stability in pitch and yaw but laterally the Saratoga could do with more self-levelling. Like so many modern aircraft, lateral stability is in short supply for reasons that are not at all clear to me. I suspect modern designers have lost the art.

Initial approach is recommended at 95 kt, which I would have thought is more than fast enough because 80 kt is adequate over the threshold, allowing speed for the round-out and a short hold-off period prior to touchdown. Like most Piper aircraft the Saratoga is easy to land.

Capabilities

Maximum ramp weight is 3,615 lb (1,640 kg) and a really well-equipped example would stand empty at 2,200 lb (998 kg). So there is a useful load of 1,415 lb (642 kg). Top up the tanks and bang goes 612 lb (278 kg) for a start. That leaves a payload of 802 lb (364 kg), which is enough for four 170-lb (77-kg) inmates and about 120 lb (54 kg) of golf clubs, etc. You could fly a with-reserve 784 nm at 159 kt (75 per cent power) or 937 nm by going for the best economy 65 per cent setting, which will do you 151 kt. Playing it the other way, six adults of average weight 160 lb (73 kg) and 180 lb (82 kg) of baggage would entail reducing the fuel load to 275 lb (125 kg), that is, 38 Imp/46 US gal (174 litres), when the safe range at 65 per cent power becomes 350 nm.

When the four seats behind the pilots are removed, there is a 4-ft wide, 4-ft high, 7-ft long cabin section, available for up to 1,100 lb (499 kg) of cargo. Naturally, the large double door comes into its own during this type of operation.

Verdict

Nearest competitor to the Saratoga SP is the Cessna Centurion (see last airtest in this chapter). Both aircraft return similar air miles per gallon, the Centurion being a little more economical at low power settings. The Saratoga has a wider cabin, which, in my view, is an advantage. Like most aircraft with long cabins relative to their overall size, the Saratoga must be loaded with care to avoid centre-of-gravity problems and the flight manual lists a fairly complex method of filling the seats and baggage areas.

For its power the Saratoga SP is not particularly fast. But it offers light-twin comfort in terms of cabin space. However, my main criticism is that it is only 6–9 kt faster in the cruise than its fixed undercarriage brother. For the extra money involved I would expect more than that and the fixed-gear version may be a better buy for many people.

Facts and figures

Dimensions

Wing span	32 ft, 2 in
Wing area	178.3 sq ft
Length	27 ft, 8 in
Height	8 ft, 6 in

Weights & loadings

Max ramp	3,615 lb (1,640 kg)
Max take-off	3,600 lb (1,633 kg)
Equipped empty	2,200 lb (998 kg)
Useful load	1,415 lb (642 kg)
Seating capacity	6/7
Max baggage	200 lb (91 kg)
Max fuel (usable)	85 Imp/102 US gal (386 litres)
Wing loading	20.2 lb/sq ft
Power loading	12 lb/hp

Performance

Max speed at sea level	164 kt
75% power cruise	159 kt
65% power cruise (best economy mixture)	151 kt
Range at 75% power	784 nm
Range at 65% power	937 nm
Rate of climb at sea level	1,010 ft/min
Take-off distance over 50 ft	1,759 ft
Landing distance over 50 ft	1,612 ft
Service ceiling	16,700 ft

Engine

Lycoming IO-540-K1G5D developing 300 hp at 2,700 rpm, driving a two-blade, 80-in (203-cm) diameter Hartzell constant-speed propeller.

TBO

2,000 hours.

Beechcraft Bonanza A36

The classic trend-setter

Background

The dust of war had barely settled before the light plane manufacturers were making ready for the promised boom in private flight. 'The skies will be full of little planes,' we were told. Well it has yet to happen, even in the United States. Among the more ambitious light singles was the Beech Bonanza. It was all metal at a time when most light planes were either fabric covered, steel tube jobs or of wooden construction; and it had what were then considered to be advanced features, such as retractable wheels and a constant-speed propeller. The prototype flew in 1945 and following exhaustive testing the first production model appeared two years later. The most striking feature was its 'V' tailplane, which made it possible to dispense with a separate fin and rudder. Because it enjoyed a large dihedral angle, the tailplane performed the functions normally handled by separate horizontal and vertical surfaces. For pitch control the elevators moved in unison but when the rudder pedals came into play the two elevators worked differentially. Why fit a 'V' tail? The obvious answer is

that by eliminating the fin and rudder you save weight and reduce drag. The original Bonanza set new standards that in many respects have never been beaten to this day. In the first year of production a staggering 1,000 had been delivered, such was the impact of this trend-setter on the aviation scene. And by 1982 more than 15,000 had been sold despite the Bonanza being the most expensive single on the market.

In 1968, 10 in (25 cm) were added to the fuselage length; they fitted an additional double door behind the trailing edge of the wing and made a four-seat club interior behind the two pilots. Engine power was increased from 260 to 285 hp, a large conventional fin and rudder were added, along with a straight tailplane, and they called it the Bonanza A36. Today there are three Bonanzas being produced by Beech, all of them powered by the 285-hp, fuel-injected Continental 10-520-BA engine. There is the Bonanza F33 with a conventional tail, the V35 (it has the

Beechcraft Bonanza A36, largest model in the Beech single-engine range of light aircraft.

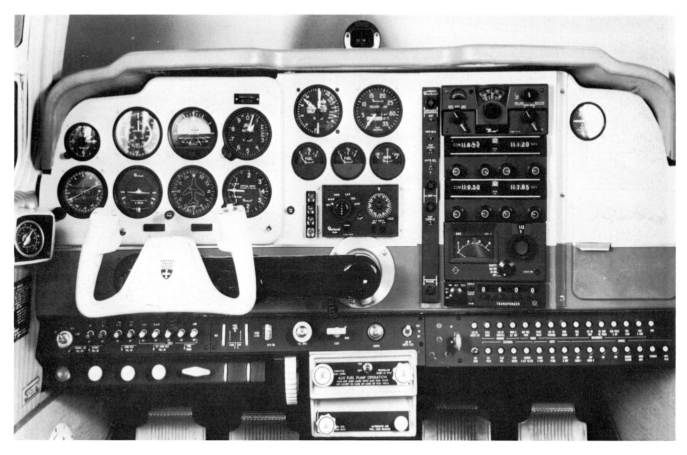

original 'V' tail and is the fastest Bonanza) and the large-fuselage A36 model described in this test report.

Engineering and design features

Seen on the ground the Bonanza A36 looks large and slightly dated but that is hardly surprising, considering the age of the design. Beech is an up-market outfit, and like all their aircraft the Bonanza reeks of quality.

The rather thick wing has leading- and trailing-edge taper. There are two mainspars and these attach to the fuselage via adjustable fittings which allow the incidence of each wing to be altered independently for correcting a persistent wing drop at the stall. In addition there are 4 degrees of washout on the wings. The flaps are a type of watered-down Fowler, which provide a modest increase in wing area. Two leading-edge tanks are buried in the wings. Standard capacity is 37 Imp/44 US gal (167 litres) usable but there is an optional 62-Imp/74-US-gal (280-litre) system. Marks inside the filler necks correspond with 100, 87½ and 75 per cent maximum fuel – a useful feature when gallons must be limited for some reason.

The tail surfaces are totally conventional, with an elevated trim tab and a small fixed surface on the rudder that is adjustable on the ground. An unusual feature for a single-engine aircraft is the spring-bias type aileron trim which is adjusted on a small knob in the centre of the control wheel.

There is a fixed step on the right-hand side of the fuselage to assist pilots onto the wing prior to entering their door at the front of the cabin. The other four occupants use the big double doors already mentioned. Four large windows are let into the cabin sides and there is a wrap-around windscreen of generous proportions.

The two cowling panels are released by quick-action catches and the usual tubular engine mountings are replaced in the Bonanza by a box-like cradle which is an extension of the main structure. The engine is canted 2½ degrees to the right for the purpose of reducing slipstream effect under varying power settings. The aircraft has a blunt nose which belies its speed.

Bonanzas are stressed to Utility Category requirements and that makes them some 15 per cent stronger than most touring aircraft. A walk around reveals more rivet heads than you would expect to find on a battleship – old-fashioned but strong.

Cabin and flight deck

The double doors open to reveal a 45-in (114-cm) wide hole giving access to the club interior (or four seats facing forward if required) and the baggage area, which is as well furnished with carpet and padding as the rest of the 'drawing room'. It will take up to 400 lb (181 kg) but you would have to watch your take-off weight before filling the tanks or the seats. The cabin is 42 in (107 cm) wide,

ABOVE: The large double door behind the wing can be seen in this air-to-air picture.

OPPOSITE: The rather dated instrument panel with its throw-over control yoke and separate sprung flight panel.

which by modern standards is a little tight. However, the interior is higher than usual – 50 in (127 cm) as opposed to the oppositon's 48 in (122 cm) – and longer than most aircraft in its class. Cabin volume is almost 130 cu ft (4 cu m), slightly more than the twin-engine Piper Seneca and 10 cu ft (0.28 cu m) more voluminous than the Cessna Centurion which has a tapered cabin with a ceiling that lowers towards the rear (see next airtest). The interior gives an impression of quiet good taste. There is a useful folding table attached to one wall. Ventilation is of the reverse-flow type with eleven fresh-air vents in the cabin – each part of the cabin may be separately heat controlled.

Up front in the pilot's end of the bomber the instrument panel and general layout can only be described as dated. Beech tend towards the conservative, being rather inclined to hang onto features long after new ones have proved themselves to be better. For example, they cling to the old 'throw-over' control wheel. When the captain has had enough of flying he lifts the control wheel on an arm pivoting from the centre of the panel and swings it in front of the pilot in the right-hand seat. All very fine. But when

you order a dual-control Bonanza with a wheel for each pilot, the lash-up – with a crossbar that intrudes into the line of vision and partly obscures some of the instruments – looks like Brooklyn Bridge.

The plunger-type throttle, mixture and propeller controls have Vernier adjustment (you twist their knobs), but for normal movement you must first depress a button in the centre of each knob – and there are times when they can be awkward. I would like to see these plungers replaced by simple levers.

The electrically operated undercarriage is controlled on a wheel-shaped switch and in the DOWN position doors close to protect the wheel wells from mud. Inside and out the Bonanza looks what it is: a Rolls-Royce among light planes. But even Rolls-Royce have been known to move with the times. The Bonanza flight deck is overdue for modernization. And they should provide a slot for stowing airways manuals.

In the air

The example flown for this airtest had the optional three-blade propeller and this jerked around before dissolving into a disc as the engine rumbled into life. Nosewheel steering is by spring linkage to the rudder pedals and radius of turn is tightened by differential use of brake. As we taxied out I was aware of the ailerons twitching in unison with the rudder pedals – those interconnecting springs again. Beech should know better than to prostitute their fine aircraft with such a corny device as this.

The ride while taxiing is excellent, visibility is above average and the brakes are among the best I have experienced on a light aircraft. I timed 12 seconds from opening the throttle to lift-off at 70 kt. There was no tendency to swing. At 100 kt rate of climb was in excess of 1,000 ft/min, which for a 285-hp tourer is not all that stimulating, but it was quiet enough to have a normal conversation, although I seem to remember the earlier 260-hp models were even quieter. At 3,500 ft, the 75 per cent power setting (2,500 rpm/24 in manifold pressure) gave an indicated 165 kt which computed into a TAS of 173 kt. That was well above the claimed performance but the aircraft was below maximum weight at the time of my flight. A fully loaded Bonanza A36 will fly a 75 per cent power cruise of 170 kt at 6,500 ft. There is a still rapid 148 kt if you climb to 12,000 ft and use 55 per cent power, a technique that adds about 11 per cent to the range.

Stalling speeds are modest enough: 63 kt clean and only 52 kt wheels and flaps down. There is a mild but adequate pre-stall buffet and the wings remained level with flaps down. Visibility is above average in most directions and the side windows curve inwards towards the roof to provide a degree of upward view. Handling is first class, partly because of good aerodynamic design but also as a result of well-engineered, friction-free control runs. The ailerons are delightful and even the spring-bias aileron trim works well. Stability is excellent in all three axes. In the air the Bonanza is, in the main, outstanding.

Being so clean and fast the aircraft could take some slowing down but for the fact that the undercarriage may be dropped at speeds of up to 150 kt. Deceleration is then quite rapid. Trim changes are slight and a rock-steady approach may be flown at 80 kt, going for 70 kt over the fence. The landing itself is so easy it's a shame to take the money.

Capabilities

The Bonanza A36 weighs in at 3,600 lb (1,633 kg). Unlike most manufacturers Beech quote a 'standard' empty weight that includes many items regarded as extras by other light-plane builders; but to bring the aircraft up to operating standards you would have to add avionics, furnishings, and so forth, weighing another 100 lb (45 kg) or so. Equipped empty weight (including the bigger fuel system) is then 2,250 lb (1,021 kg), leaving a useful load of 1,350 lb (612 kg). Filling the tanks will subtract 444 lb (201 kg) from that, allowing a payload of 906 lb (411 kg) – enough for five 170-pounders (77-kg people) and 56 lb (25 kg) of baggage or four males of average weight 160 lb (73 kg) plus two females each of 133 lb (60 kg). When need arises to carry the usual airline baggage allowance of 44 lb (20 kg) each you are limited to four occupants.

With full tanks the Bonanza A36 will fly a with-reserve 700 nm at 170 kt or 750 nm when the 65 per cent power setting is used to provide a speed of 162 kt. The more

recently designed Aerospatiale Trinidad (see page 86) will fly a similar cabin load, 2 kt faster for another 160 nm – and on 35 less hp. Six people of average weight and their baggage – 200 lb (91 kg) each – entails reducing fuel uplift in the Bonanza to 150 lb (68 kg), ie, 21 Imp/25 US gal (95 litres), when the range at 170 kt becomes 236 nm.

Verdict

In terms of payload/range the Bonanza A36 does not compare well with the Cessna Centurion but it can match the Piper Saratoga. It is not as fast as the Centurion but more rapid than the Saratoga, although the latter has the widest cabin. Above all else the Bonanza is among the best handling of the light singles and is in a class of its own for high-quality engineering. Almost 80 per cent of all Bonanza models built since 1947 are still flying. Surely that speaks for itself.

Facts and figures

Dimensions

Wing span	33 ft, 6 in
Wing area	181 sq ft
Length	27 ft, 6 in
Height	8 ft, 5 in

Weights & loadings

Max take-off	3,600 lb (1,633 kg)
Equipped empty	2,250 lb (1,021 kg)
Useful load	1,350 lb (612 kg)
Seating capacity	6
Max baggage	400 lb (181 kg)
Max fuel (usable)	62 Imp/74 gal (280 litres)
Wing loading	19.9 lb/sq ft
Power loading	12.6 lb/hp

Performance

Max speed at sea level	178 kt
75% power cruise at 6,500 ft	170 kt
65% power cruise at 10,000 ft	162 kt
Range at 75% power	700 nm
Range at 65% power	750 nm
Rate of climb at sea level	1,015 ft/min
Take-off distance over 50 ft	2,165 ft
Landing distance over 50 ft	1,575 ft
Service ceiling	16,000 ft

Engine

Continental 10-520-BA developing 285 hp at 2,700 rpm, driving an 84-in (213-cm) diameter McCauley two-blade propeller or an optional 80-in (203-cm) diameter, three-blade propeller.

TBO

1,800 hours.

Cessna P210N Pressurized Centurion II

A step towards all-weather singles

Background

There are a lot of Cessna light singles about and some airfields abound in these high-wing birds: 152 trainers and 172 tourers with, here and there, a more powerful Skylane. But if ever you spot one bigger and more muscular looking than the rest, with a three-blade windshovel on the front, odds are that it is the model 210 Centurion.

Cessna have been building 'two-tens' since 1960 but the early models bear little resemblance to the current ones.

The first Cessna 210 was powered by a 260-hp Continental engine but four years later engine output was increased to 285 hp and a new wing was added, although, like the first model 210, it was strut braced. Then in 1967 there were big changes. The wing became laminar flow, there was a straight leading edge, with a trailing edge providing constant taper from root to tip, and the struts disappeared. Maximum weight went up by 700 lb (318 kg), although power remained the same. However, all this cleaning up added about 7 kt to the cruise despite the gain in weight.

The Piper Saratoga described previously is offered in four versions: a fixed gear model, the retractable SP (described in the book), and turbocharged versions of both. In many respects the Cessna 210 Centurion is similar in category but all models are retractable. There is a normally aspirated version powered by a 300-hp Continental, the Turbo Centurion, and the Pressurized Centurion (tested for this report) which at time of publication is the only single-engine, pressurized aircraft in production.

Engineering and design features

The wing, which as already mentioned is of totally different planform to other Cessnas, is of similar area to the much lighter Skyhawk (see Chapter 3). It sports down-swept wing-tips that are intended to minimize induced drag, and there are the usual neat and tidy Cessna, Fowler-type flaps. Cessna have replaced those dreadful flat-plate affairs with excellent Frise ailerons and, being a cantilever structure, this is a very clean wing, completely unspoiled by the usual brackets some manufacturers insist on adding here and there. Two wing tanks provide a total usable capacity of 74 Imp/89 US gal (337 litres) and the fuel selector may be set to LEFT, RIGHT and OFF, but not 'both' tanks. The fuel pump switch is a 'split' type: the right half (coloured yellow) is used for starting and continuous operation in the event of mechanical pump failure; the left-hand segment is an emergency switch for when maximum fuel flow is required, say, during the initial climb. Appropriately it is coloured red but I feel that

Cessna P210N Pressurized Centurion II. The 'bomb' under the right wing contains the scanner for the weather radar.

Cessna should replace this complicated system with something that does not require of the pilot a PhD in engineering.

The tail surfaces are comprised of a fixed tailplane and separate elevators with a large trim tab on the right-hand trailing edge. There is a generously proportioned fin which starts from behind the cabin as a dorsal area and sweeps up to meet the rudder hinge line.

The massive fuselage looks bigger than it is because the large cabin windows of the unpressurized Centurion models have been replaced by small ones that look like square portholes. The cabin is pressurized to +3.35 psi and the big windows have had to go. On the face of it this does not appear to be much of a pressure differential but it nevertheless ensures a cabin altitude of just over 12,000 ft when the aircraft is at its maximum operating level of 23,000 ft – and that is very worthwhile.

All six occupants must enter and leave through a large single door on the left, although there is a hinge-up panel on the right which doubles as an emergency exit and a source of fresh air while on the ground in hot weather. The Continental TSIO-520-P engine develops 310 hp for take-off and has a maximum continuous rating of 285 hp. It drives an 80-in (203-cm) diameter, three-blade propeller.

The underside of the fuselage has a slightly pregnant look where the mainwheels disappear when retracted. One way or another the main legs go through some pretty involved gymnastics, dangling straight down and twisting before finally vanishing from view. Before this mind-blowing performance the wheel bay doors open hydraulically and close to hide the undercarriage afterwards.

The example that I flew was equipped with weather radar, essential in any aircraft capable of climbing through 20,000 ft of cloud where CBs might be growing, and had the scanner housed in a streamlined 'bomb' hanging from a small pylon under the right-hand wing. It also had de-icing boots on the flying surfaces.

Seen on the ground the Pressurized Centurion is an impressive beast and must be one of the cleanest high-wing designs of all-time.

Cabin and flight deck

Although there is only one door, entry to the cabin is not unduly difficult, even for those occupying the rear bench seat. At its widest point the cabin is 44 in (112 cm) across but it tapers somewhat towards the rear. It is about 5 in (13 cm) narrower than the Saratoga, 6 in (15 cm) narrower than the Trinidad but it has a 2-in (5-cm) advantage over the Beech Bonanza (see previous tests) and has the longest cabin in its class. Behind the rearmost seats is a baggage hold with its own separate door where up to 200 lb (91 kg) may be loaded. Apart from the rear bench already mentioned there is a pair of seats for a pilot and co-pilot (or passenger) followed by another pair making a total of six in all. As an option you can have 'all-dancing' pilots' seats that not only adjust for leg reach, height and angle of back rake but also provide pneumatic lumber support which contours itself to you and your slipped discs at the press of a button – quite the best seats I have seen. There is enough leg room for all occupants.

Sitting up front you could easily imagine that this is an executive jet or turboprop. The deep instrument panel carries three levels of instruments on the left; a large avionics/autopilot stack in the centre; with, to the right, the engine instruments, which are angled towards the pilot, and the weather radar. Engine controls (throttle, pitch and mixture) take the form of plungers – a pity, considering the professional standard of everything else on this remarkable aircraft. Furthermore, there is no proper stowage space for airways manuals.

Running down to the floor is a central console with the elevator and rudder trim wheels, an engine cooling flap lever, a microphone and the fuel selector. Below the captain's control wheel is a panel for the pressurization dump valve (you operate that just before landing to ensure that no cabin pressure remains and it is safe to open the door without becoming a circus act), the master switch, a bank of rocker switches for the electric services, the undercarriage lever and its position lights (the flap lever is on the other side of the engine controls) and the simple pressurization management controls. Other than a few warning lights, there is no comprehensive annunciator, which is, perhaps, surprising.

Before starting the engine you close the pressure dump valve, move the pressurization switch to ON and set the cabin altitude selector to approximately 1,000 ft above the departure or arrival airfield, whichever is the higher. It is as simple as that and pressurization will begin as the aircraft climbs through the cabin altitude set on the selector. Here are a few numbers for you to ponder:

Aircraft altitude (ft)	Cabin altitude (ft)
7,000	Sea level
10,000	2,400
14,000	5,500
18,000	8,500
23,000	12,100

Pressure for the cabin is taken from the turbocharger and it therefore follows that you should not close the throttle suddenly when flying above, say, 10,000 ft, particularly when any of the occupants suffer from sensitive ears. An emergency oxygen system is fitted and there is an excellent cabin-heat/windscreen-defrost facility with outlets at floor and ceiling level. The pilot can monitor the behaviour of his pressurization system on a cabin rate-of-climb indicator and a cabin altitude/differential pressure indicator. If at any time the cabin altitude exceeds 12,400 ft a red warning light comes on, oxygen masks should then be worn or a descent to lower levels be made immediately.

The cabin is comfortable and I liked the flight deck layout. But the Pressurized Centurion suffers from the usual Cessna problem of having a high-set instrument panel which, in their singles, obstructs the view ahead unless you have the optional height-adjusting seat. Further-

The high-set instrument panel provides adequate room for radio, weather radar (top right) and the pressurization controls (bottom left).

more, the pressurized version suffers more restricted visibility than the other Centurion models. There are thick door pillars and the passengers must make do with very much smaller windows. However, this is a pressurized aircraft and you must expect limitations of this kind.

In the air
The aircraft is easy to taxi, with good brakes, accurate nosewheel steering and a comfortable ride. If you feel happy taking off in a Cessna 152 (see Chapter 2), then you will like the Pressurized Centurion, which is just as easy, only steadier. Rotate speed is a modest 65–70 kt and the undercarriage retracts in about 11 seconds. There is a 5-minute limit for use of full power but it is surely good practice to set up climbing power at an early opportunity.

You can expect a climb rate of 930 ft/min for a fully loaded example; mine went up at around 1,300 ft/min but I estimate we were 700 lb (318 kg) below the 4,000-lb (1,814-kg) maximum at the time. For its power this is not a particularly good sea-level climb, however, being turbocharged, the P210N will almost maintain its best rate of climb up to 10,000 ft and even at 20,000 ft will go up at more than 500 ft/min.

At Flight Level 160 (16,000 ft altitude on a standard day), 74 per cent power (2,300 rpm/33 in manifold pressure) produced an indicated 150 kt which, according to my high-class navigation computer, trued out at a rapid 190 kt. There was no fuss and the aircraft was very quiet; pressurization tends to push the noise out of the cabin rather than let it in. Under these conditions fuel burn is 14.6 Imp/17 US gal (64.4 litres) per hour or, to keep the motorists happy, a fully loaded example could fly six people at 220 mph and return over 15 miles per Imperial gallon. Maximum cruising speed is attained at 20,000 ft where 79 per cent power should provide you with 195 kt.

I thought that handling was not up to the very high standards of the nonpressurized Centurions but there is more control-run friction in this model, possibly because of the need to provide pressure seals where the cables depart the cabin. Nevertheless, and in spite of an interconnecting aileron-to-rudder spring, I thought that handling was above average. Visibility in flight is not up to the standard of unpressurized models and Cessna should take steps to improve it. They could start by lowering the top of the instrument panel.

The stall with full-flap came at an indicated 55 kt to the accompaniment of a sharp left-wing drop. Cessna should do something about that – it is quite unnecessary in this day and age. I thought the electric elevator trim was painfully slow but apparently the United States authorities (FAA) insist on this, which is a pity. There is moderate but nevertheless positive lateral stability and plenty of it in pitch and yaw. The aircraft has that sure, confident feel you experience in a light twin.

Final approach is flown at 70–80 kt, according to weight, and they don't come any easier to land than the Centurion.

Capabilities

With a ramp weight of 4,016 lb (1,822 kg) the Pressurized Centurion is a giant among light singles. An example equipped to average standards would have a useful load of 1,604 lb (728 kg). Full tanks will take 534 lb (242 kg) of that and leave a generous 1,070 lb (485 kg) for pilot and passengers. Even if all on board weighed 170 lb (77 kg) a head there would be enough left for 50 lb (23 kg) of baggage. More usually there will be some people in the aircraft who weigh, say, 140 lb (64 kg) or less when the baggage allowance would increase pro rata. With full tanks, range will depend on flight level and power setting. Here are a few 'for instances' taken from the flight manual:

Altitude (ft)	% Power	TAS (kt)	Range (nm)
20,000	71	187	815
16,000	50	145	910
12,000	60	160	870
10,000	71	170	780

By any standards the Cessna Pressurized Centurion is quite a performer. Even with a full cabin and 200 lb of baggage you can still fly a with-reserve 595 nm at almost 190 kt.

Verdict

The Pressurized Centurion is one of the most expensive light singles on the market, costing more than the smallest twins. However, it does enable you to fly over the weather without needing to wear oxygen masks and without the discomfort of high flying in a low-pressure cabin. It is faster than some of the quite powerful light twins, many of which it will out-range, while returning very much better air miles per gallon.

If most of your flying is over land and away from large mountain areas or other inhospitable terrain then this aircraft with its economic, over-the-weather capabilities, is probably a better buy than a light twin. Personally, I think it is an outstanding plane in most respects

Facts and figures

Dimensions

Wing span	36 ft, 9 in
Wing area	175 sq ft
Length	28 ft, 2 in
Height	9 ft, 5 in

Weights & loadings

Max ramp	4,016 lb (1,822 kg)
Max take-off	4,000 lb (1,814 kg)
Equipped empty	2,412 lb (1,094 kg)
Useful load	1,604 lb (728 kg)
Seating capacity	6
Max baggage	200 lb (91 kg)
Max fuel (usable)	74 Imp/89 US gal (337 litres)
Wing loading	22.9 lb/sq ft
Power loading	12.9 lb/hp

Performance

Max speed at 17,000 ft	206 kt
79% power cruise at 20,000 ft	195 kt
60% power cruise at 12,000 ft	160 kt
Range at 79% power at 20,000 ft	770 nm
Range at 60% power at 12,000 ft	870 nm
Rate of climb at sea level	930 ft/min
Take-off distance over 50 ft	2,160 ft
Landing distance over 50 ft	1,500 ft
Service ceiling	23,000 ft

Engine

Continental TSIO-520-P developing 310 hp at 2,700 rpm (5-minute take-off rating), 285 hp at 2,600 rpm (max continuous rating), driving an 80-in (203-cm) diameter, three-blade, constant-speed propeller.

TBO

1,800 hours.

Some close formation by the author. Note the small cabin windows which replace the larger ones of un-pressurized Centurions.

HOW DO THEY RATE?

The following assessments compare aircraft of similar class. For example, four stars (Above average) means when rated against other designs of the same type.

★★★★★ Exceptional
★★★★ Above average
★★★ Average
★★ Below average
★ Unacceptable

Aircraft type	Appearance	Engineering	Comfort	Noise level	Visibility	Handling	Payload/ Range	Field Performance	Cruising Speed	Economy	Value for Money
Mooney M20J 201	★★	★★★★	★★★	★★★	★★	★★★★	★★★★	★★★	★★★★	★★★★★	★★★★
Aerospatiale TB20 Trinidad	★★★★	★★★★	★★★★★	★★★	★★★★	★★★★	★★★★	★★★★	★★★★	★★★★	★★★★
Siai Marchetti SF260	★★★★★	★★★★★	★★★	★★★	★★★★★	★★★★★	★★★★	★★★	★★★★★	★★★★	★★
Piper PA-32R-301 Saratoga SP	★★★	★★★	★★★★★	★★★★	★★★	★★★	★★★	★★★	★★★	★★★	★★★
Beechcraft Bonanza A36	★★★★	★★★★★	★★★	★★★★	★★★★	★★★★★	★★	★★★	★★★★	★★★	★★★
Cessna P210N Pressurized Centurion II	★★★	★★★★	★★★	★★★★	★★	★★★★	★★★★★	★★★	★★★★	★★★★	★★★

5. PISTON-ENGINE TWINS

Why fit more than one engine in an aircraft? After all, as a cynical old fighter pilot mate of mine was fond of saying: 'Put two fans on the thing and you have twice as much to go wrong.' In the early days, around 1914, the urge to add more engines was motivated by a need for more power than was available from one motor. Even so, rates of climb with all props turning was by modern standards pathetic. Some of the early twins would stagger up at only 300–400 ft/min and when an engine failed there was no question of continuing the flight because your canvas and wire birdcage went down on one fan quicker than it would go up on two.

Between the World Wars, as engine power increased, serious thought was given to the charms of multi-engine or even twin-engine safety. One or two of the pre-war twins behaved quite well on one engine but it was not until after the Second World War that more stringent requirements were enacted for civil aircraft. Those intended for public transport had to be capable of climbing with a full load following engine failure during or soon after take-off. Less stringent regulations apply to light twins, but it is a bad design that will not provide at least a token rate of climb on one engine. So to come back to the original question, why fit more than one engine? The short answer is safety.

To a considerable extent the need for more than one engine diminishes as they become more reliable. These days it is not difficult to provide enough power in one unit and there are obvious economic advantages to installing, say, one 500-hp motor rather than two, each developing 250 hp. However, the decision as to whether you should buy one of the powerful singles described in the previous chapter or go for the added security of two engines must depend upon the routes you will be flying most often. If you are likely to spend most of your time over the flat, populated lands of Europe or North America, where adequate communications exist, then the more advanced light single would probably give good service. On the other hand, frequent water crossings, repeated flights over forests or mountains or long trips across sparsely inhabited areas are not places where failure of the one and only engine can be accepted philosophically.

Although few of the singles are fitted with ice protection even the smallest twins are often to be seen with quite good systems, and these can greatly enhance the value of an aircraft in parts of the world where icing conditions are an everyday fact of aviation life. When it is required

to fly more than six or seven people then, with one possible exception, the light twin becomes essential.

What, then, is expected of a good twin-engine, piston-powered aircraft?

- Bearing in mind the high initial and running costs, it must offer a reasonable rate of climb when, at maximum weight, an engine quits soon after lift-off. Many light twins fail miserably on this count, largely because certifying authorities in the countries concerned do not have the stomach to demand suitable requirements. Indeed, the manufacturers would rightly claim that, however miserable the engine-out rate of climb might be, their wonderplanes exceed the laid-down requirements. I would regard 400 ft/min on one engine as good, 300 ft/min acceptable and 250 ft/min tolerable. But anything less has got to be bad news, because on a hot, bumpy day a pilot of only average ability may be unable to produce a rate of climb on one engine at a time when he most needs it.
- Standards of comfort should be higher than in a single.
- Noise levels must be lower than in a single.
- The aircraft must be easy to fly. A twin that becomes a juggling act when it has to be flown on one engine has no place in modern aviation.
- It must be easy to fly manually, particularly on instruments. There have been a few relatively unstable twins on offer in recent times and the favourite excuse put forward by the manufacturers is: 'We expect our airplane to be flown on the autopilot.' Take my advice: any salesman who tells you that should be shown the door.
- While accepting that even a low-powered twin cannot match the economy of a high-powered single, air miles per gallon remain important and the ideal twin should offer good seat miles per pound of fuel.
- Landing weight should equal or approach maximum take-off weight. You should not have to burn off an hour's fuel before landing back at base, when, for any reason, an unscheduled return is demanded.

Before describing the piston-engine twins in this chapter I would like to assure readers who have never flown behind more than one engine that you do not have to be a superman to make the conversion to twin-engine aircraft. But the consequences of bad airmanship can be more serious than in a single.

Engine failure in a single is a black-and-white situation. There is no power, and gravity claims its debt. In a twin the loss of one engine should allow the pilot to continue flying safely provided the airspeed is sufficient to ensure adequate rudder control. That is essential if direction is to be maintained against the pull of the live engine. This is called minimum control speed.

Because minimum control speed varies according to aircraft configuration (wheels up or down, flaps-up or in the landing position) and other considerations such as aircraft weight, a list of speed classifications, known as the V Code, has evolved. Many of these are purely academic, while others may be regarded as an exercise in hair splitting. Some only relate to large aircraft; the following list is therefore limited to those I will be mentioning in the airtest reports.

Single-engine capabilities spell safety in a twin. Beechcraft Pressurised Baron.

V Code

V_1 Decision speed during take-off. Up to that speed you should have enough runway in which to stop if an engine fails. Beyond that speed you must take off, because there is not enough runway ahead.

V_r Rotate speed. Speed at which the nose is lifted to attain the take-off attitude.

V_2 Take-off safety speed. This is Minimum Control Speed with a safety margin added to cater for:
1. Element of surprise;
2. Failure of the critical engine;
3. Landing gear down, flaps in the take-off position, propeller windmilling on the failed engine;
4. Pilot of average strength and ability.

V_{mc} Minimum control speed. Minimum speed at which it is possible to maintain direction following failure of the critical engine (that is, the one that causes most yaw when the other has failed). There are several of these. V_{mca} relates to in the air; V_{mcg} applies to on the ground before lift off; and V_{mcl} refers to minimum control speed in the landing configuration.

V_{ne} Never exceed speed. The ASI will be marked with a red line at that speed.

V_y Speed for best rate of climb on all engines.

V_{yse} Speed for best single-engine rate of climb, often marked on the ASI as a blue line and sometimes known as blue-line speed.

V_{no} Normal operating speed. Sometimes called maximum structural cruising speed. Beyond it we enter the yellow arc on the ASI. This must be avoided in turbulence.

V_{at} Target threshold speed for 'over the fence' during final approach.

Piper PA-44-180T Turbo Seminole

The twin-engine Warrior with a 'T' tail

Background

First and smallest of the post-war light twins was the British Miles Gemini, a delightful little wooden four-seater, brought out in 1946, with an electrically retracting undercarriage and 900 miles range on a pair of 100-hp Cirrus Minor engines. With so little power it could only maintain height on one engine at light weights. A little larger, but not much better on one engine, was the original Piper Apache which, in its later guise as the Piper Aztec, received a pair of 250-hp engines and other, relatively minor changes. Piper had a sleek single-engine job, the Comanche and it was not long before the Apache was replaced by a twin-engine version of this – the Twin Comanche. It was fast, quiet, very efficient but no respector of fools, so you had to be in no doubt about the asymmetric exercises in the event of an engine failure. Around 2,150 of these fine little twins were built before a disastrous flood at the plant washed away all jigs. It was the end of the road for the Twin Comanche.

Some years were to pass before any manufacturer ventured into the miniature twin market. Then in 1977 Grumman in America were first in the field with a potentially good design known as the Cougar. It flew on the same power as the Twin Comanche, two 160-hp engines, but Grumman lost interest in light aviation and ceased production of their entire range of singles and the Cougar.

Soon afterwards Beech brought out the Duchess (see next airtest), and Piper introduced the Seminole. At 180 hp a side, both aircraft are more powerful than the Cougar. The subject of this airtest is the turbocharged version of the Seminole.

Engineering and design features

Piper are masters of building new planes out of bits of this and that. The Seminole follows that tradition in so far as much of the fuselage, wing, etc, is from the single-engine Warrior (see Chapter 3). The tapered wings have a 24-in (61-cm) extension outboard of the ailerons, otherwise they are very similar to those of its single-engine relatives. So are the mechanically operated, slotted flaps which have the usual settings that remain unchanged from the Cherokee of many years ago – UP, 10 degrees, 25 degrees (maximum lift position) and 40 degrees for landing.

All fuel is carried in tanks positioned in the rear of the engine nacells. Total usable capacity is 90 Imp/108 US gal (409 litres). The engines counter-rotate; seen from behind, the left one goes around clockwise while its brother on the right turns anti-clockwise. The result? – well, there is no critical engine in the event of engine failure and the usual slipstream/torque effects which try to swing the aircraft during take-off or whenever the throttles are adjusted is cancelled by the opposing rotation. The Lycoming TO-360 engines are nicely cowled and turbochargers maintain their rated 180 hp from sea level to 12,000 ft. To cater for a wide range of temperature conditions engine cooling flaps are fitted. Piper claim you can completely remove an engine cowling in five minutes. An unusual and very useful feature of the aircraft is the provision of accurate tubular dipsticks in the filler cap of each tank and a convenient single-point fuel strainer mounted behind a small door in the side of the fuselage adjacent to the wing.

The fibreglass nose hinges up to allow easy maintenance of the electro-hydraulic power pack, battery and combustion heater. Undercarriage retraction takes only six seconds and although wheels-up must be selected below 110 kt the gear can be lowered at 140 kt, which is useful for providing additional drag while entering the circuit. Since the wheels are held up by hydraulic pressure, failure of the system results in their dropping down and locking under the influence of gravity.

The large fin has a dorsal area, the rudder is endowed with one of the biggest trim tabs I have seen on a little bomber and the stabilator is mounted almost at the top of the vertical surfaces. Whatever else this arrangement does for the Seminole it spoils the look of an otherwise graceful little twin. The fuselage has three windows on each side and a large, two-piece windscreen. There is a 22 × 20 in (56 × 25 cm) door for the 24 cu ft (0.68 cu m) baggage area where up to 200 lb (91 kg) may be carried. A single entrance door is provided on the right-hand side of the fuselage. The Seminole may lack the Twin Comanche's grace and aura of speed but it is nevertheless attractive when seen on the ground.

Cabin and flight deck

The door is closed easily with its long lever which pulls it into the weather seals without having to slam. At just under 42 in (107 cm) wide the cabin is more than 2 in (5 cm) narrower than its principal rival, the Beech Duchess. However, Piper have made the most of what space there is, and design, trim and finish of the example tested reflected the high standards now being attained by Piper.

The seats are firm but comfortable and height adjustment for the pilots is available as an option. Headrests are provided for all four seats. I was glad to see that Piper have resisted the temptation of turning a good four-seater into a thoroughly bad six-seat aircraft, something they did with the Twin Comanche.

Instrument layout follows closely that described for the Warrior (see Chapter 3). There is room for two vertical stacks of avionics and the demonstration aircraft had weather radar which I would have thought ostentatious for an unpressurized aircraft, although admittedly there may be some parts of the world where it is needed. Below the main panel is a strip containing five rectangular instruments for each engine: temperatures, pressures, fuel contents, and so on. There is also an engine gas temperature (EGT) gauge for fine adjustment of the mixture. It has a left-engine/right-engine selector switch. Nearby are the autopilot controls, the undercarriage selector with its warning lights, and the engine controls (pairs of throttle, mixture and propeller levers), which all reside in a quadrant. Battery master switch and others for the electric services are located on a small panel, along with the ignition and starter switches. These are positioned under the captain's left window. The starter takes the form of a left-engine/right-engine rocker switch with an electric priming button adjacent to each arm.

I liked the cabin and thought the flight deck area was well planned.

In the air

Starting arrangements for the turbocharged, carburettor-type Lycoming engines are standard except for the electric priming buttons, which are convenient and effective. Visibility while taxiing is very good and so are the brakes but the nosewheel steering feels spongy and I suspect the spring linkage should be beefed up a little. The turbocharger system is a simple one in the interest of low cost and there

is no automatic boost control. An overboost warning light illuminates at 36.1 in manifold pressure (36.5 in must not be exceeded) and it helps to set take-off power against the brakes before starting the roll. V_r (rotate speed) is 70 kt and in the event of engine failure blue-line speed is marked at 88 kt. Take-off is simple, the Turbo Seminole getting itself off the ground and at the 50-ft point in about the same distance as its single-engine cousin, the Warrior. Initial rate of climb was 1,400 ft/min and this we maintained up to 10,000 ft. Fully loaded Piper claim 1,290 ft/min.

At Flight Level 100 (10,000 ft on a standard day), 75 per cent power returned a TAS of 174 kt, 5 kt faster than book value for a fully loaded example. If you are prepared to wear oxygen masks and cruise at 18,000 ft, speeds of up to 183 kt are attainable using 75 per cent power. Noise levels are lower than average at all power settings.

The Turbo Seminole handles very much better than the average jelly-on-wings and Piper have gone to town in the stability department which is good in pitch, roll and yaw. Furthermore, you can trim the aircraft at, say, 140 kt, lower the undercarriage and operate the mechanical flaps without trim changes, showing what can be done when the flight development boys try hard enough. Certainly the aircraft is a pleasure to fly even on one engine.

However, a fully loaded Turbo Seminole has a miserable single-engine climb rate of only 180 ft/min and that is by no means good enough. On a hot and perhaps turbulent

Piper Turbo Seminole, a typical example of this manufacturer's talent for using existing components to create a new aircraft.

day a pilot of limited experience would find it difficult to climb safely if he lost an engine soon after take-off. In all other respects, including the stall at 55 kt with full-flap and the landing (which is simplicity itself), the Seminole is a fine little aircraft.

Capabilities

Ramp weight for the Turbo Seminole is 3,943 lb (1,789 kg) and an airways equipped example would have an empty weight of 2,560 lb (1,161 kg), allowing 1,383 lb (627 kg) of useful load. Four 170-lb (77-kg) adults each with 44 lb (20 kg) of baggage would represent a payload of 856 lb (388 kg), leaving 527 lb (239 kg) for fuel, which represents 73 Imp/88 US gal (333 litres). That would be enough to fly a with-reserve 530 nm at 75 per cent power (170-kt TAS) or 550 nm at 65 per cent (160 kt). In each case the altitude is 12,000 ft and while higher speeds are available if you are prepared to cruise at up to 18,000 ft, that would mean wearing oxygen masks. The ranges allow for the usual start-up, taxi, power checks, climb and a 45-minute diversion.

With full tanks payload is reduced to 735 lb (333 kg), say, four 170-lb (77-kg) adults, and 55 lb (25 kg) of baggage between them or two men at 170 lb (77 kg) and two ladies weighing a more becoming 130 lb (59 kg) along with 34 lb (15 kg) of baggage each. Range is then 650 nm at 75 per cent power and 680 nm at 65 per cent. For the sake of a few knots these can be increased to 740 nm and 770 nm respectively when the economy mixture setting is used.

Verdict

While the Turbo Seminole is a good performer, to obtain full advantage from it the aircraft must be flown at cruising levels where oxygen masks have to be worn. Not everyone takes kindly to the experience and there is the added risk of ear damage to occupants with a cold. I know you are not supposed to fly with a head cold but people do these things.

The turbocharged version of the Seminole weighs 125 lb (58 kg) more than its normally aspirated brother and that brings with it a hidden disadvantage. Maximum landing weight is 3,800 lb (1,724 kg) and you could find yourself flying around for almost an hour to burn off fuel if, for any reason, it was necessary to return and land

following a maximum weight take-off. The Beech Duchess is cleared to land at its maximum weight. Why not the Seminole?

In terms of fuel economy the Turbo Seminole will return 7.2–7.5 nm/Imp gal (6–6.3 nm/US gal or 1.58–1.65 nm/litre), but the Beech Duchess can better that by about 11 per cent.

Unless you have to operate from hot and high airfields I suspect the normally aspirated Seminole is a more practical aircraft. Up to 12,000 ft there is not much difference in cruising speed, it has a better rate of climb and it is cheaper to buy. Must say, I liked the quiet cabin.

Facts and figures

Dimensions

Wing span	36 ft, 7 in
Wing area	183.8 sq ft
Length	27 ft, 7 in
Height	8 ft, 6 in

Weights & loadings

Max ramp	3,943 lb (1,789 kg)
Max take-off	3,925 lb (1,780 kg)
Max landing	3,800 lb (1,724 kg)
Equipped empty	2,560 lb (1,161 kg)
Useful load	1,383 lb (627 kg)
Seating capacity	4
Max baggage	200 lb (91 kg)
Max fuel	90 Imp/108 US gal (409 litres)
Wing loading	21.4 lb/sq ft
Power loading	10.9 lb/hp

Performance

Max speed	195 kt
75% power cruise at 12,000 ft	170 kt
65% power cruise at 12,000 ft	160 kt
Range at 75% power (12,000 ft)	650 nm
Range at 65% power (12,000 ft)	680 nm
Rate of climb at sea level	
Two engines	1,290 ft/min
One engine	180 ft/min
Take-off distance over 50 ft	1,500 ft
Landing distance over 50 ft	1,400 ft
Landing distance with optional heavy-duty brakes	1,190 ft
Service ceiling	
Two engines	over 20,000 ft
One engine	12,500 ft

Engines

2 × Lycoming TO-360-E1A6D developing 180 hp at 2,575 rpm, driving two-blade, Hartzell feathering propellers.

TBO

1,800 hours.

OPPOSITE ABOVE: The flight deck is not particularly wide but Piper have used the available space to good effect.

OPPOSITE BELOW: Clean installation of the two engines is evident in this air-to-air picture.

Beechcraft Duchess

A little twin with a lot of style

Background
Soon after the now-discontinued Grumman Cougar light twin appeared as the first of the Twin Comanche replacements, Piper brought out the Seminole (described in the previous test). During 1978 Beech introduced their Duchess. Like the Piper offering, Beech have based their little twin on one of their single-engine designs, the Sierra. There is nothing particularly new about making twins out of existing singles. In 1946 the Miles Co of Reading, England, devised the Gemini from their single-engine Messenger. There is a lot to be said for the technique, after all, it makes sense to use existing, well-proven components or structures – provided they are ideal for the new design.

Other than their excellent Bonanza (see Chapter 4) I have never been a fan of Beech singles. I thought their Sport 150 was an expensive and over-engineered Cherokee, while the higher-powered Beech Sundowner cost considerably more than a Piper Archer, yet had less useful load, a low cruising speed and very poor air miles per gallon. Furthermore, the Beech Sierra retractable sported the sort of cabin in which inmates could wear top hats – there was that much headroom. As a result it was a disappointing performer for a 200-hp retractable. But its twin-engine development, the Duchess, is a different ball game and in this book it rates highly.

Engineering and design features
Surprisingly, Beech have elected to use the parallel-chord wing of the Sierra rather than the very efficient tapered one of their Bonanza. Slotted flaps are carried on six rather large and clumsy external brackets, otherwise the structure is free of blemishes. A bonded fuel tank is located in the front portion of each wing. Total usable capacity is 83 Imp/100 US gal (379 litres). Filler points, which reside near the wing-tips, have visual check marks at 60, 80 and 100 per cent fuel – a useful aid while taking on the correct amount of fuel for the trip. Overload a twin and there will be no single-engine performance if an engine fails.

Carburettor-type, 180-hp Lycoming engines are installed within close-fitting cowlings. These are very similar to the motors used in the normally aspirated version of the Piper Seminole. The undercarriage is retracted and lowered by a self-contained electro-hydraulic power pack supported by a mechanical emergency lowering system. The now trendy 'T' tail has a fixed tailplane with separate elevators, which resides almost at the top of a very large fin and rudder.

The fuselage is based upon the single-engine Sierra and it benefits in looks from a long, well-proportioned nose which slopes down to ensure good forward visibility from the cabin. Two doors are provided, left and right of the front seats, and there is another one almost as large on the left, leading to the 200-lb (91-kg) capacity baggage area at the back of the cabin.

G-BGHP, the plane tested, was of particular interest because it was the first little twin to be fitted with TKS ice protection. The TKS system utilizes thin, porous steel strips applied to the wing, tailplane and fin leading edges through which de-icing fluid is metered when the need arises. The advantages to this system are that it does not affect the aircraft's performance like rubber boots. And when the weather is such that icing is not on the menu no fluid need be carried, so the system is lighter. Also it lasts indefinitely, not like boots, which must be replaced every four to five years. Equally important, it demands less skill of the pilot to operate effectively; pneumatic boots can be mismanaged and cause problems. On the debit side the TKS system costs more to install although this is offset by its long life and the fact that rubber boots can cost more to replace after four or five years than the original installation. And when fluid is carried to deal with expected icing conditions the TKS system is heavier than pneumatic boots although here again, this is partly or even fully compensated for by the fuel economy that results from the undiminished cruising performance of aircraft using fluid de-icing. On balance I must prefer TKS to the various rubber-boot systems, which are now old-fashioned. They belong to the age of steam.

As a matter of personal taste I prefer the looks of the Duchess to the Seminole's. The former has more style, although I wish Beech had fitted a tapered wing. A parallel-chord mainplane cannot be efficient however hard it tries. There is little to choose in terms of engineering quality between the two aircraft.

Cabin and flight deck
Thanks to the large doors entry and exit is easy and the 44-in(112-cm) wide cabin offers a little more elbow room than the Seminole. Four individual seats are provided and the example flown was upholstered in a rather fetching red plush material of a kind that was fashionable in Victorian days. The deep instrument panel, which carries three rows of instruments and a full-width electric-switch and circuit-breaker strip along the bottom of the main area, is nevertheless set low enough to allow unobstructed vision ahead.

The flight panel and radionav readouts are positioned squarely in front of the captain with the engine instruments to their right, which occupy the centre of the panel. The right-hand half of the panel is devoted to avionics in two stacks, the autopilot control box and space for a second altimeter. Controlling the ice protection is a neat panel that contains push buttons, warning lights and a fluid contents gauge. It may be set in the anti-icing mode (to prevent airframe ice forming) or de-icing (to remove it when it has formed). Three hours worth of fluid is carried – more than enough because normally you only need protection during the climb and descent, and the system is cleared for flight into known icing conditions.

The flap switch is a hateful affair which must first be pulled out before it can be moved. And if you want to set, say, 15 degrees, the switch must be returned to the OFF position to stop the flaps at that setting. If Cessna can give us a 'follow-up' system (the only one fit for a pilot with a lot on his mind) on their cheap little 152 (see Chapter 2) then Beech should wake up to the present and do the same for their excellent little twin.

There is a power quadrant in the centre of the panel with the usual pairs of levers for the throttles, mixture controls and the propeller pitch, along with carburettor heat controls, elevator and rudder trim wheels and a small knob that provides aileron trim via spring pressure. A combustion heater is located in the nose. It takes about 2 gal (9 litres)/hour out of the right-hand tank. This is a splendid cabin and the flight deck is very convenient and easy to use.

In the air

Engine starting is managed on a panel just above the captain's left knee. The combined starter/ignition switches are of the rotary type. Handling on the ground is first class with good nosewheel steering and powerful brakes. Visibility is above average ahead and to the sides but in the air the roof obscures the pilot's vision to the right, particularly when banking in that direction, during a turn, for example.

For the flight we were some 500 lb (227 kg) below maximum take-off weight. V_{mca} was 71 kt, acceleration was rapid and lift-off followed a few seconds after passing that speed. Wheels- and flaps-up, and settled at an IAS of 85 kt, the initial rate of climb was 1,350 ft/min. The strange thing about Beech is they do not give power settings for 75 per cent, or for that matter any other per cent, which is a pity. Instead the manual talks in terms of 'maximum cruise power', 'recommended cruise power' and 'economy cruise power'. Tables of figures are provided under each heading giving IAS and TAS against cruising altitude with separate columns of figures for ISA−20°C, standard day (ISA) and ISA+20°C. And having refused to mention power percentages, the manufacturers go on to advise pilots not to exceed 75 per cent power until the oil consumption of new engines stabilizes. So work that out if you can!

It so happens that the Duchess advertising brochure tells us that 75 per cent power ' . . . zips you along at 166 kt (191 mph).' Relating this to the flight manual I was able

Beechcraft Duchess, similar in concept to Piper's Seminole (see previous test).

Although the instrument panel is deep, view ahead is good.

The nose sweeps down and provides excellent forward visibility.

to determine that if you set 2,700 rpm and 24 in manifold pressure at 6,000 ft you will have 75 per cent power and a cruising speed of 166 kt. At 7,000 ft the 'recommended cruise power' (2,500 rpm and 24 in manifold pressure) provided a TAS of 163 kt, a lower-than-average noise level and a total fuel burn of 16.94 Imp/20.2 US gal (64 litres) per hour. Handling is excellent, with crisp, well harmonized controls. If you shut down an engine it is easy to maintain direction on the ailerons alone while the feet are removed from the rudder pedals. Rudder trim is adequate for returning to balanced flight on one engine and without increasing power on the live one there was a TAS of 116 kt. I managed to get the speed down to 65 kt IAS before running out of rudder against full power on the live engine. The propellers, by the way, counter-rotate.

At 7,000 ft the single-engine rate of climb was 350 ft/min but a fully loaded Duchess is claimed to go up at only 235 ft/min when a donkey gives up at sea level — better than the Turbo Seminole but not good enough. Stability in all three axes is excellent and during stalling the Duchess behaved like one, showing no tendency to drop a wing. Wheels- and flaps-up the stall came at 60 kt IAS and full-flap (30 degrees) brought the 'g' break in at 53 kt. The Duchess is a delightful little aircraft to fly and you rapidly feel at home within its hospitable surroundings.

Initial approach at 85 kt followed by 75 kt over the fence worked well. There was, however, a lot of turbulence on the way in and I thought the ailerons required coarse use to deal with the conditions. Perhaps they are a little too low geared in relation to control wheel movement. Trim changes are slight and even in the prevailing cross-wind and turbulent conditions the landing did not demand the touch of an ace.

Capabilities
Maximum ramp weight is 3,916 lb (1,776 kg) and the well-equipped example tested had an empty weight of 2,600 lb (1,179 kg) which included the ice protection system. This leaves a useful load of 1,316 lb (597 kg). Full tanks would account for 598 lb (271 kg) and leave a payload of 718 lb (326 kg). Four 170-lb (77-kg) occupants could take 38 lb (17 kg) of baggage between them, but when any of them were lighter the baggage load could be increased on a pound for pound basis. Range with maximum fuel (with reserves) is 680 nm at Beech's so-called 'maximum cruise power', 710 nm at 'recommended cruise power' and 750 nm when the economy setting is used. A full cabin of four 170-pounders and 200 lb (91 kg) of baggage would entail reducing fuel by 25 Imp/30 US gal (114 litres) when the range becomes 475–520 nm, according to power setting. The Turbo Seminole will fly that load for another 100 nm although there is not a lot to choose in terms of full-tank range.

Verdict
In terms of handling, the Duchess is even better than the Seminole. It also has the advantage of not suffering from a landing weight restriction. Although it has a slightly wider cabin (wide enough to make a difference to passenger comfort) the Duchess matches the cruising speed and maximum range of the Seminole (normally aspirated version) on the same engine power.

You can always point a finger at some feature that, in the view of the critic, is below standard — like the flap switch. But other than its single-engine climb performance I thought the Duchess was an above-average little aircraft in most respects. And the example tested with its anti-icing system can be regarded as a serious all-weather aircraft. Indeed the British agents have used it with considerable success on air taxi work throughout Europe on a year-round basis. It is good value too.

Facts and figures

Dimensions
Wing span	38 ft
Wing area	181 sq ft
Length	29 ft
Height	9 ft, 6 in

Weights & loadings
Max ramp	3,916 lb (1,776 kg)
Max take-off	3,900 lb (1,769 kg)
Equipped empty	2,600 lb (1,179 kg)
Useful load	1,316 lb (597 kg)
Seating capacity	4
Max baggage	200 lb (597 kg)
Max fuel	83 Imp/100 US gal (379 litres)
Wing loading	21.5 lb/sq ft
Power loading	10.8 lb/hp

Performance
Max speed at sea level	171 kt
'High-speed cruise'	166 kt
'Recommended cruise'	162 kt
Range at 'high-speed cruise'	680 nm
Range at 'recommended cruise'	710 nm
Max range at 143 kt	850 nm
Rate of climb at sea level	
Two engines	1,248 ft/min
One engine	235 ft/min
Take-off distance over 50 ft	2,120 ft
Landing distance over 50 ft	1,880 ft
Service ceiling	
Two engines	19,650 ft
One engine	6,170 ft

Engines
2 × Lycoming 0-360-A1G6D developing 180 hp at 2,700 rpm, driving Hartzell 76-in (193-cm) diameter, two-blade propellers.

TBO
2,000 hours.

Partenavia P68C

Sleek simplicity from Naples

Background
Luigi Pascale is not only Professor of Aeronautical Engineering at Naples University, he is also a keen sporting pilot taking an active part in the Italian light-aviation scene. Light planes of his design have consistently won his country's major event, the *Aero Internazionale D'Italia*, for as long as anyone can remember and this was recognized in 1982 when he was presented with a trophy.

The first of his light-plane designs to become translated into hardware was the P48 Astore, which he built in a small lock-up workshop near his home. Several designs later the Italian government selected one of his aircraft for the flying schools and a small factory was set up at Arzano near Naples. Thus began the Partenavia Co. They produced several hundred four-seat tourers, all of them sold within Italy, but soon Professor Pascale had plans for the P68, a simply constructed light twin of advanced design. The P68 is a six/seven-seat, high-wing design powered by a pair of 200-hp Lycoming engines. The logistics of carting through the streets wings, fuselages, tail surfaces and all the bits and pieces for assembly at Naples Airport was entertaining for the locals of Arzano but a pain in the unmentionable to Partenavia. So on 19 January,

1974 they upped sticks and moved into a fine new factory just across the road from Naples Airport. The P68 sells at a steady rate and by 1982 about 250 of them had been delivered to customers all over the world. There is a turbocharged version, a fisheries patrol/spotter development with a clear plastic nose and a stretched one powered by Allison turboprop engines. The example featured in this test report is the normally aspirated P68C.

Engineering and design features
The aim of the design is to transport a pilot and up to six passengers as economically as possible but not at the expense of performance. Economy is a function of good air miles per gallon, low depreciation charges, reliability and cheap maintenance, the last three conditions being achieved by this aircraft through simplicity of design.

The parallel-chord, cantilever wing has electrically operated slotted flaps, Frise ailerons and a high-aspect ratio. Two wing tanks provide a total usable fuel capacity of 114 Imp/137 US gal (519 litres). There is a swept fin and

Partenavia P68C. The graceful lines of the fuselage hide its roomy cabin.

Flight deck layout is conventional. This example has weather radar (bottom of the radio stack).

rudder, allied to a small dorsal area and an all-flying tail-plane (stabilator).

A pair of Lycoming engines is slung under the wings within neat, close-fitting cowlings and when ice protection is specified the wing leading edges outboard of the nacelles are fitted with pneumatic boots, as are the fin and stabil-ator leading edges.

The fuselage is brilliantly designed and, like any vehicle of near-perfect proportions, it contrives to be bigger inside than seems possible when viewed from outside. A gently sloping nose continues in contour with the windscreen then gradually flattens to join the upper surface of the wing, from where it descends almost imperceptibly towards the tail. The under-surface of the fuselage is another unbroken sweep, blending with the overall design to create an impression of engineering as an art. Fuselage drag must be very low.

There are three large windows on each side of the cabin, the passenger/crew door is on the left and another large door is situated behind the passenger areas on the right. It leads to a voluminous baggage area with a capacity of 400 lb (181 kg). For parachute training this door may be removed to make an ideal exit for those with a mind to 'get out and walk'.

The mainwheels are supported on fixed spring steel legs and there is a telescopic nose strut. It may seem surprising that a fixed undercarriage should have been chosen for such a clean airframe but simplicity was uppermost in Pascale's mind. In any case they tried retractable gear on the slightly faster turboprop version and gained a miser-able 6 kt in cruising speed – three minutes saved on a 250-nm journey at a cost of more complexity, higher initial and maintenance charges and a reduction in useful load caused by the addition of a hydraulic system.

Seen on the ground the Partenavia P68C is a beautiful bird. And as I said earlier, it is hard to credit that seven people can sit in its cabin, enjoying reasonable standards of comfort.

Cabin and flight deck

At 44.5 in (115.5 cm) wide the cabin provides more elbow room than some of the more powerful high-performance twins. Firm but comfortable seating is arranged with two individual chairs for the pilot and front passenger, another pair of seats behind and a bench at the rear of the cabin for three people. A 6-in (15-cm) gap between the individual seats provides a small aisle, allowing you to move through the cabin. There are individual fresh-air vents and reading lights for each passenger; and I would describe the cabin as functional rather than luxurious.

The pilot – or when required, pilots – are provided with a well-designed flight deck. All engine instruments are on the extreme right, the avionics and the radionav readouts take up the centre of the panel and the flight instruments are laid out in standard form before the pilot. Most of the electric switches reside along the bottom of the left-hand panel and the circuit breakers are on the wall below the captain's side window.

Two fuel selectors in the roof provide normal supply and crossfeed in a simple and straightforward manner. Early examples of the P68 had a bizarre arrangement straight out of *Alice through the Looking Glass*, and some dramatic moments are on record when pilots frantically twisted knobs the other way to restore power following an uninvited silence.

There is a power management quadrant full of throttle, mixture and pitch levers, along with elevator and rudder trim wheels and their position indicators. Early models of the P68 had a rather high glareshield, but either they have lowered this or raised the height of the seat because the model 'C' provided better forward visibility than the one I flew some years previously. It is a neat, tidy, logical 'office' that is easy to work in.

In the air

The engines are fuel injected and you start them accordingly: prime with the electric pump, return the mixture control to the cut-off position, activate the starter and then move forward the mixture control as she fires. I thought the fuel flow indicators were not all that easy to read from the left-hand seat.

Taxiing is a push-over. The brakes and nosewheel steering are first class and any Cessna 152 captain would feel at ease in this little twin. One of the special talents of these aircraft is their remarkable field performance and rate of climb. V_{mca} (minimum control speed, air) is only 62 kt, and blue-line speed for best rate of climb in the event of an engine failure is 88 kt. I opened the throttles, we accelerated rather noisily and left the ground in less than 800 ft. The aircraft was fully loaded at the time but we went up at 1,600 ft/min. At 5,000 ft, 70 per cent power (2,300 rpm/24 in manifold pressure) gave a TAS of 154 kt

Clean entry is ensured by careful blending of the nose and windscreen profiles.

for an economic total fuel burn of only 16 Imp/19.2 US gal (72.8 litres) per hour.

One of the problems shared by all high-wing propeller-driven multis or twins is noise. The under-surface of the mainplane acts as a sound-board which directs engine and propeller noise into the cabin. The Partenavia P68C is by no means the noisiest I have flown but it could do with an intensive study aimed at making the cabin quieter. Double windows, better sound insulating materials and perhaps a rearrangement of the exhaust system could work wonders. Also, the existing four-cylinder engines might be replaced by quieter six-cylinder units of similar power. Piper did this with the developed models of their Seneca (see the next test) and the reduction in cabin noise was significant.

Handling is excellent in the Italian tradition and so is stability, particularly in pitch: only one cycle was required to regain a trimmed speed after I had raised the nose to induce a 15-kt displacement. The example tested had an excellent electric elevator trim and an effective manual one for the rudder. Without flap the stall came, wings level, at 68 kt, and full-flap (35 degrees) made a worthwhile 13-kt reduction at the 'g' break.

Although the P68C has a fixed undercarriage it offers a better single-engine rate of climb than the more powerful Piper Seneca III and the now-discontinued Piper Aztec, which enjoyed the advantage of disappearing wheels and another 100 hp.

Trim changes are minimal when flap is lowered and you aim for 80 kt on short finals. The aircraft sits low on the ground and this must be borne in mind when the round-out is made prior to touchdown. Remember that and the P68C will be very easy to land.

Capabilities

Maximum take-off weight is 4,387 lb (1,990 kg) and a typically equipped example would provide its proud owner with a useful load of around 1,400 lb (635 kg) – according to the weight of equipment carried. Six people, each with 40 lb (18 kg) of baggage would represent a payload of 1,200 lb (544 kg) and leave 200 lb (91 kg) for fuel, that is, 28 Imp/33 US gal (125 litres). Allowing for taxi, run-up, take-off, climb and a 45-minute reserve for possible diversions, safe range with such a load (which I would regard as the maximum for comfort, although seven people can be carried) is about 180 nm. Of course if the pilot was doing an air taxi job and had no need of his 40 lb (18 kg) of baggage and if some of the passengers turn the scales at less than 160–170 lb (73–77 kg), fuel load goes up *pro rata* and ranges of 250–350 nm become possible with six people.

Full tanks allow a payload of 579 lb (263 kg) – a pilot, say, and two passengers with 44 lb (20 kg) of baggage each – and then ranges of 1,000–1,200 nm are possible although the aircraft is really intended as a short-distance, cheap-to-operate light twin. In terms of field performance the P68C will operate out of airstrips that would entail emptying the cabin of other light twins.

Verdict

The Partenavia P68C is a step up in size from the two four-seat miniature twins previously described (Piper Seminole and Beech Duchess), although with its fixed undercarriage it is a slightly simpler piece of engineering. Its nearest competitor in the 200-hp-a-side class is the Piper Seneca III, which is powered by 220-hp turbocharged engines and which has a retractable undercarriage. However, unlike the Seneca, Partenavia's little airliner can be landed at its maximum take-off weight. Also, the Partenavia P68C is very much cheaper to buy and operate, and over short distances is almost as fast on the journey. It is all a matter of horses for courses, but either way the P68C is a great little aircraft.

Facts and figures

Dimensions

Wing span	39 ft, 5 in
Wing area	200 sq ft
Length	31 ft, 4 in
Height	11 ft, 2 in

Weights & loadings

Max take-off	4,387 lb (1,990 kg)
Equipped empty	2,987 lb (1,355 kg)
Useful load	1,400 lb (635 kg)
Seating capacity	6/7
Max baggage	400 lb (181 kg)
Max fuel	114 Imp/137 US gal (519 litres)
Wing loading	21.91 lb/sq ft
Power loading	10.97 lb/hp

Performance

Max speed at sea level	174 kt
75% power cruise at 7,500 ft	166 kt
65% power cruise at 11,000 ft	161 kt
Range at 75% power	1,050 nm
Range at 65% power	1,140 nm
Max range at 150 kt	1,210 nm
Rate of climb at sea level	
Two engines	1,500 ft/min
One engine	280 ft/min
Take-off distance over 50 ft	1,300 ft
Landing distance over 50 ft	1,600 ft
Service ceiling	
Two engines	19,200 ft
One engine	6,900 ft

Engines

2 × Lycoming IO-360-A1B6 developing 200 hp at 2,700 rpm, driving Hartzell two-blade propellers.

TBO

2,000 hours.

Piper PA-34-220T Seneca III

General aviation's best-selling twin

Background

First of Piper's wide-body light singles was the Cherokee Six. It was most things I dislike in an aircraft, not a pretty sight, an undistinguished performer for its power and, in terms of handling, a milkshake on wings. It did, however, have a massive, comfortable cabin and it was not long before Piper were offering a twin-engine retractable version which they called the Seneca. I did not think much of that either. It was noisy, handling was dreadful and it had some of the worst trim changes I have ever experienced in a light aircraft. In the late 1960s Piper were going through a bad patch in terms of handling but Lynn Helms became boss man and pretty soon they were taking on professional test pilots who began to inject a little taste into the operation.

Within a few years Piper had taken their Seneca apart. They replaced the insipid flat plates on the wings with a pair of Frise ailerons, the 200-hp Lycoming engines – four-cylinder jobs that are good, reliable but noisy – were swapped for six-cylinder, turbocharged Continentals

which maintained their power up to 12,000 ft, these and various other improvements resulting in the Seneca II. Then in 1981 a still more developed version, the Seneca III, was introduced. Engine rpm have been increased from 2,600 to 2,800, power has gone up by 10 hp to 220 hp and this can be maintained up to 15,000 ft where there is a 20-hp-a-side increase over the Seneca II. The instrument panel has also been given a face-lift. Whatever I may have thought of the early Seneca, and my views were shared by others, the breed has proved to be a best-seller among light twins, partly because it is good value but no doubt its big cabin has a lot to do with the fact that more Senecas are sold than all five models of the Beech Baron combined (see airtest later in this chapter). It even outsells Piper's single-

Piper Seneca III, a twin-engine development of the Cherokee 6/Saratoga singles that has become the best-selling aircraft in its class.

The massive passenger door with its adjacent opening comes into its own when the aircraft is used as a light freighter.

engine Warrior! (See Chapter 3.) But you should not really be too surprised, because the Seneca III is the right size at the right price.

Engineering and design features
The fuselage is basically that of the single-engine Saratoga (see Chapter 4), which itself is a development of the Cherokee Six. To provide a big cabin of more or less constant cross-section the underside is flat until several feet behind the wing trailing edge, then it sweeps up sharply towards the tail surfaces. There is a large double door at the back of the cabin on the left and another for the crew up front on the right. Two baggage doors are provided, one in the nose and another behind the cabin area. Each baggage hold will take up to 100 lb (45 kg).

Four generous-size windows are let into the fuselage sides and there is a new, one-piece windscreen. In place of the Saratoga's engine compartment is a long, shapely nose which contrives to turn what is really an ugly box into a fuselage of quite elegant aspect when seen from more flattering angles. At the blunt end of the ship is a very large swept fin, a horn-balanced rudder and a small stabilator (all-flying tailplane to use the English translation) with the usual combined anti-balance/trim tab. The wing has no taper except for the leading edges between the fuselage and the engine nacelles. It carries the Frise ailerons already

mentioned and Piper's mechanically operated slotted flaps which are a straight pinch from the original Cherokee.

There are two fuel tanks in each wing but only two filler points. Standard capacity is 78 Imp/94 US gal (356 litres) but there is an optional 102-Imp/123-US-gal (466-litre) system which I would strongly recommend. The wings have 7 degrees of dihedral, the minimum in my opinion for satisfactory lateral stability. Undercarriage raising and lowering is by reversible electro-hydraulic power pack; it goes around one way for UP and does an about turn for DOWN. Only seven seconds are required to perform the vanishing trick and although the maximum speed for retraction is a not very fast 108 kt they may be lowered at up to 130 kt – useful for reducing speed while joining circuit traffic.

The two counter-rotating Continental engines reside within smart-looking cowlings and although two-blade propellers are standard fit you can spend a lot more money – and give away 38 lb (17 kg) in useful load – by installing three blades: ostentatious for 220-hp engines.

The Seneca III may be something of a box on wings but from some angles it looks positively sleek.

Cabin and flight deck

Entry and exit is easy for pilot and passengers alike and I was immediately impressed by the long, 49-in (125-cm) wide cabin which lends itself to various interior layouts. The example flown by me had two seats up front for a pilot and passenger (or two pilots if required). Another rear-facing pair backs onto these and there are two forward-facing seats at the far end of the cabin. A small table can be set up between the facing pairs, and seating is very comfortable for crew and passengers alike. There is plenty of leg-room all round and such is the width of the cabin that a useful refreshment unit may be installed between the middle pair of seats. There are fresh-air vents and reading lights for all occupants as well as built-in oxygen points for everyone. Those would be needed to make full use of the aircraft's performance advantages which become particularly attractive when cruising above 14,000 ft.

Early models of the Seneca had their control yokes offset slightly but now Piper have contrived to position them where they should have been in the first place — in front of the pilots. The very wide instrument panel has room for two radio stacks, one in the centre (above the power quadrant) and another to the right. All circuit breakers are low down on the right and, if required for crew training purposes, a second flight panel may be installed for an instructor in the right-hand seat.

Engine instruments are to the left of centre and electric switches have been removed from their original location on the left wall to the main panel above the captain's left knee. The autopilot controller is on the far left and the wheel-shaped undercarriage switch, together with the usual warning lights, is just to the left of the power quadrant.

A central console incorporates the power quadrant with its throttle, mixture and propeller controls in pairs of levers for the left and right engines. Below them are alternate air selectors (for use when the normal engine induction intakes are affected by ice) and, moving back between the pilots, a fuel selector for each engine marked ON, OFF and X FEED for use in an engine-fail situation when, for example, the live engine on the left needs to draw fuel off the right-hand wing tanks.

Rudder and elevator trim wheels reside nearby along with their position indicators. Between the two front seats is a large, pull-up lever for the flaps. Other than the mechanical flaps, about which I shall say more later, I thought the Seneca III flight deck was intelligently planned. The example flown was graced by one of Piper's de-luxe interiors — all crushed velour and front stalls with champagne to follow. Very impressive. The finish of Piper aircraft, inside and out, has come a long way since the 1960s.

In the air

As an option you can fit electric priming buttons. These are installed next to each arm of the LEFT/RIGHT rocker switch that starts both engines. The nosewheel steers via the rudder pedals through 27 degrees left and right of centre and the effort required is reasonable. Visibility could be better ahead but it is good to the sides. Piper should consider lowering the glareshield an inch or two. There are plenty of normal pilots on both sides of the Atlantic. Not all of us stand 6 ft, 6 in and have to bend double while entering a room.

For this airtest the aircraft was within 275 lb (125 kg) of its maximum take-off weight. This is a price-conscious twin and the turbochargers have no frills; consequently you must exercise care not to exceed 40 in manifold pressure. 'Overboost' warning lights will illuminate at 39.8 in. To save watching the lights during take-off Piper recommend setting up power on the brakes. V_{mca} is 66 kt and blue-line speed for best single-engine rate of climb is 92 kt. For a normal take-off no flap is used.

I set up power on the brakes, released them and the machinery accelerated smoothly to 80 kt when a little back pressure on the control yoke lifted us cleanly off the runway. Initial climb rate at 92 kt was around 2,000 ft/min, which is considerably in excess of the claimed 1,400 ft/min for a fully loaded aircraft. Eventually we settled at 1,600 ft/min, a climb rate that was maintained up to 10,000 ft. 75 per cent power at that altitude (2,500 rpm/33 in manifold pressure) resulted in a TAS of 180 kt, and 65 per cent (2,400 rpm/32 in manifold pressure) reduced the speed by only 3 kt. Fastest 75 per cent cruise mentioned in the Seneca flight manual is 193 kt but you would have to climb to 16,000 ft and that would entail wearing oxygen masks.

I would say that the Seneca III is a little quieter than average but not much. Certainly there is a big improvement on the earlier models in this respect.

Flaps-up there was a wings-level stall at 63 kt and this reduced to a full-flap (40 degrees) stall at 53 kt. General handling is very much better than the earlier models but the ailerons are not all that effective, and the stabilator feels like . . . well, a stabilator. It is heavy and rather spongy. Likewise the elevator trim is heavy to use and the throttles must be adjusted slowly because of the time needed for the turbochargers to settle at the new manifold pressure selected. Power management without tears requires a little practice in the Seneca III.

Lateral stability is better than average but two complete cycles were need to regain a trimmed IAS of 160 kt after a 15-kt displacement. Asymmetric flight is perfectly straightforward but the single-engine climb rate of 240 ft/min is marginal, although not the worst I have experienced. Without adding power on the live engine there was a single-engine TAS of 135 kt at 8,000 ft.

Initial approach is flown at 90 kt and I experienced some difficulty in getting down full flap. I would have thought that the Seneca is a little on the big side for manually operated flaps and time is long overdue for electric ones. Few women pilots would be able to apply full-flap with one hand. Over the fence, 75 kt is ideal and the aircraft is simple to land.

Capabilities

The Seneca III was given twenty-eight modifications from the previous model. Some of these were quite minor, the more important ones I have already mentioned. Maximum ramp weight is 4,773 lb (2,165 kg) and a well-equipped example would provide a useful load of around 1,550 lb (703 kg). With full tanks (long-range version) there is a payload of 815 lb (370 kg), which will do nicely for four adults, each weighing 170 lb (77 kg), and 135 lb (61 kg) of baggage. Assuming we avoid using oxygen and cruise at 10,000–12,000 ft such a load can be flown over a with-reserve 640 nm at 75 per cent power, while 65 per cent would fly you at 175 kt and add 120 nm to the range. A full cabin – 6 adults, each weighing 170 lb (77 kg), plus 200 lb (91 kg) of baggage – would entail reducing fuel by

46 Imp/55 US gal (208 litres), the range at 75 per cent power becoming 340 nm. A cabin of mixed-weight occupants and reduced baggage could probably be flown over a with-reserve 450 nm. As a freighter, a 170-lb (77-kg) pilot could fly a 1,070-lb (485-kg) load for 300 nm.

The 4,513-lb (2,047-kg) maximum landing weight is a disadvantage. You would have to fly around for more than an hour to burn off fuel if, for any reason, it was necessary to return after a maximum-weight departure.

Verdict
The Seneca III may best be described as a good all-rounder: reasonably fast, reasonably economical, reasonable in terms of payload/range and more reliable than the earlier models which tended to wear out parts a little too quickly. One of the Seneca's biggest assets is its roomy cabin. But Piper really must improve the trimmer and the flap system – some of us are getting too old for weight-lifting. And there is no excuse for its landing-weight limitation.

From some angles the rather boxy Seneca is not without grace.

Facts and figures

Dimensions

Wing span	38 ft, 11 in
Wing area	208.7 sq ft
Length	28 ft, 7 in
Height	9 ft, 11 in

Weights & loadings

Max ramp	4,773 lb (2,165 kg)
Max take-off	4,750 lb (2,155 kg)
Max landing	4,513 lb (2,047 kg)
Equipped empty	3,223 lb (1,462 kg)
Useful load	1,550 lb (703 kg)
Seating capacity	6/7
Max baggage	200 lb (91 kg)
Max fuel (long-range system)	102 Imp/123 US gal (466 litres)
Wing loading	22.8 lb/sq ft
Power loading	10.8 lb/hp

Performance

Max Speed at 14,000 ft	196 kt
75% power cruise at 10,000 ft	178 kt
65% power cruise at 10,000 ft	175 kt
Range at 75% power	640 nm
Range at 65% power	760 nm
Rate of climb at sea level	
Two engines	1,400 ft/min
One engine	240 ft/min
Take-off distance over 50 ft	1,210 ft
Landing distance over 50 ft	1,978 ft
Service ceiling	
Two engines	25,000 ft
One engine	12,300 ft

Engines
2 × Continental TSIO-360-KB developing 220 hp at 2,800 rpm, driving Hartzell 76-in (193-cm) diameter, two-blade propellers.

TBO
1,800 hours.

Cessna T303 Crusader

A cabin-class twin in miniature

Background

When the Twin Comanche replacement bug took a hold at Grumman, Piper and Beech, it was expected that Cessna would be among the feverish. Then word got around that Cessna were having second thoughts about miniature twins. They had the Skymaster with its 'centreline thrust', where an engine was positioned at each end of the cabin, the aim being to produce a twin that could be flown by a single-engine pilot with little further training. Then there was their model 310, one of the early post-war light twins which started life as something of a hot rod but developed into a marvellous plane in most respects. The Skymaster was not my idea of an aircraft but the Cessna 310 had a lot of charm. It should go down in history as one of aviation's greats.

Cessna decided to replace both the Skymaster and the 310 by a new model. While I do not lament the passing of their push-pull contraption, anything that followed the 310 would have to be good, in fact very, very good. When it emerged, the Cessna Crusader proved to be another masterpiece.

If I am critical of Cessna's lower-powered singles, my admiration for their light twins is unqualified. And in my opinion the Crusader shines, even when viewed against the background of larger and more powerful members of the Cessna family. At 250 hp a side the Crusader is considerably more powerful than the Piper Seminole or Beech Duchess previously described, but it has only 30 more horses in each nacelle than the Piper Seneca III (see previous airtest). Yet somehow Cessna have contrived to make it one of the smallest cabin-class aircraft in the business. The more powerful Cessna 310 (285-hp engines) was an 'over-the-wing-and-enter' aircraft like the Beech Baron (see test report later in this chapter) and the now-discontinued Piper Aztec. The Crusader has a set of airstairs which let down to allow airliner-type entry in style.

Engineering and design features

The Crusader is an entirely fresh design, not a collection of bits and pieces from previous models. The wing is slightly tapered and it has Fowler-type flaps similar to the very neat ones fitted on their single-engine range. Airflow regulators in the form of perforated strips extend with the flaps where they pass under the engine nacelles. Their purpose is to continue the airflow in these critical areas. Well-designed Frise ailerons are provided and small slats are fixed at the junctions between the wing leading edges and the fuselage/engine nacelle sides. These delay the breakdown of airflow at high angles of attack and reduce inter-

ference drag. In the Crusader airframe there has been a move away from plastic; even the wing-tips are metal.

Each wing has an integral fuel tank and total usable capacity is 127 Imp/153 US gal (579 litres). Power is supplied by a pair of counter-rotating fuel-injected Continental engines fitted with Ai Research turbochargers which boost them very slightly to 32.5 in manifold pressure. This is little more than standard sea-level pressure but the main object is to maintain the engine's 250 hp up to 15,000 ft or more. Being only moderately supercharged the engines enjoy a recommended TBO of 2,000 hours – a longer life than any other turbocharged motor. Unusual in a light twin, engine fire-warning systems are standard equipment on the Crusader – well done, Cessna! Engine cowlings are held in place by the usual quarter-turn, quick-release screws and the panels may be removed in minutes.

Built into the rear of each engine nacelle is a luggage locker with a capacity of 120 lb (54 kg). There is a 150-lb (68-kg) baggage hold in the nose and another 200 lb (91 kg) may be stowed at the rear of the cabin. This adds up to 590 lb (267 kg) but you should not get too carried away by such a massive baggage allowance. Take that out of the useful load, strap in six people and you can only have enough fuel for about twenty-five minutes flying. But whoever heard of six people in any aircraft wanting to fly with almost 100 lb (45 kg) of baggage *each*?

Three-bladed propellers are fitted as standard and they have cut one engine control for the pilot to worry about by replacing the cowl flaps with cooling louvres cut in the sides of the nacelles.

Running in a straight line from above the entrance to the swept fin is a dorsal area. On the side of the rear fuselage is a small handle that twists to lock the rudder against gust damage on the ground. If you forget to unlock it during the walk-around it automatically springs out of action when the control wheel is moved aft of neutral for lift-off. In any case you would be unable to steer on the ground with the rudder locked.

The tailplane is located quarter way up the fin where it not only looks better than those 'T' tails but also manages to keep itself out of the slipstream. While some may feel the Crusader is a little chunky compared with the very sleek model 310 it replaces, the fuselage is nevertheless beautifully proportioned. There are four big cabin windows on each side, a large, two-piece windscreen and a perfectly matching nose; it all hangs nicely together. The lever-suspension-type undercarriage is retracted by an electro-hydraulic power pack. These days engine-driven hydraulic pumps seem to be in a minority in light aircraft.

On the left-hand side, behind the wing, is an entrance door, the top portion hinging up to become a canopy and help keep the customers dry while entering on a rainy day, the lower section dropping down to form a set of airstairs. Viewed with the entrance open the Crusader looks every inch the miniature airliner. It is a most appealing, Disney-like creation.

Cabin and flight deck
At 48 in (122 cm) wide and 49 in (125 cm) high the Crusader's cabin is fractionally smaller in cross-section than the Piper Seneca's although it is several feet longer, but by clever design Cessna have managed to create the impression that this is a walk-around passenger area. And so it is if you stand 49 in (125 cm) high. However, an 8-in (20-cm) central aisle adds to the illusion of spaciousness and allows you to enter a seat with relative ease. The seats, by the way, have folding arm rests, head supports and individual amenity panels containing a reading light, fresh-air vents and, when fitted, oxygen point. Seating in the example flown was arranged club fashion behind the two pilots, that is, two pairs facing, and standards of comfort were very high.

Joining with the passenger area is the flight deck. The crew seats are easy to reach and the control layout follows a pattern that has become more or less standard on the Cessna twins. On the left side wall is a neat switch panel for the electric services, with a large one below it containing the circuit breakers. Ignition, battery master and alternator switches reside on a strip below the control wheel. Nearby is the undercarriage selector and the warning lights. Maximum lowering speed is a very useful 175 kt.

To the right of centre is the follow-up flap switch with its position indicator and on the extreme right of the strip panel are the cabin environment controls, a 45,000 BTU heater being fitted as standard. There is a small toggle under the captain's control yoke for applying the parking brakes.

The manifold pressure gauges, rpm indicators and fuel flow indicator take the form of three, dual finger (left engine and right engine) dials in a row along the top of the main panel. To their left is an altitude alerter and the annunciator panel. Radio is in two stacks, one down the centre, the other to its right. Flight instruments and the radionav readouts are in front of the captain.

In the centre, between the two front seats, is a power console fit for an airliner. The two throttles and the two propeller levers all had black knobs which I think is a mistake. Anything to avoid confusion (that is why outside France mixture levers are red). Mr Cessna, please colour your pitch lever blue. They provide you with a large elevator trim wheel and two smaller ones for the rudder and the ailerons. Controls for the very simple fuel system are at the base of the pedestal. There is a selector for each engine which can be moved to OFF, ON or CROSSFEED.

You would be very hard to please if this flight deck did not earn your approval.

In the air
The entrance door was closed and the DOOR OPEN warning light extinguished on the annunciator. With full

Cessna T303 Crusader, first of the cabin-class light twins.

The passenger area contrives to look larger than its measurements. Another pair of passenger seats faces forward. This picture was taken from one of these.

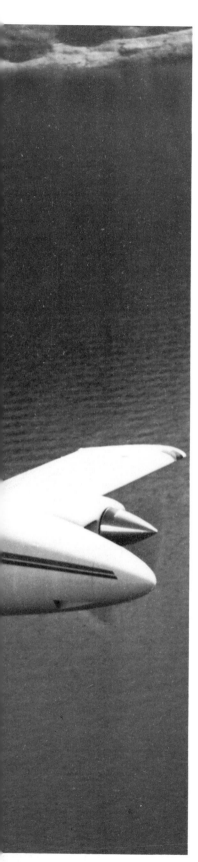

The very clean engine installation is shown to advantage.

fuel on board and four others joining me the aircraft was at high average weight. Fired up, the fuel-injected engines idled sweetly and we moved off. Visibility on the ground is excellent and nosewheel steering is lighter than some of the singles. Flap 10 degrees is used for take-off.

Initial acceleration on opening the throttles was purposeful. I checked we had not exceeded 32.5 in manifold pressure and at 75 kt eased back on the wheel, lifted off, touched the brakes to stop the wheels spinning and then retracted the gear. Best rate of climb speed is 103 kt when you can count on a climb rate of around 1,500 ft/min at maximum weight. A more practical technique is the cruise/climb which provides more than 1,000 ft/min at a sensible 120-kt IAS.

At 8,000 ft, with the rate of climb undiminished by kind permission of those turbochargers, 73 per cent power (2,400 rpm/24 in manifold pressure) gave a TAS of 176 kt and 61 per cent (2,300 rpm/22 in manifold pressure) turned in 169 kt. Both figures were up on book value.

Apart from their push-pull Skymaster Cessna have always excelled in producing twins that handle like thoroughbreds, and the Crusader is typical of their art in this respect. It could almost be a small jet, with ailerons as smooth as silk and nicely harmonized controls. The electric elevator trim was first class but the rudder trim is in need of a little more muscle; it would not return the ball to the centre when I shut down an engine.

Stability in pitch and yaw is excellent but there is not much lateral stability which is a pity in such an outstanding aircraft.

Wheels down and with maximum flap (30 degrees) the stall was preceded by pronounced elevator buffet before a gentle 'g' break at 55 kt, wings level. There must be a lot of position error at high angles of attack because Cessna quote 62 kt at maximum weight. Asymmetric flying makes few demands on the pilot but the maximum-weight single-engine climb of only 220 ft/min is disappointing and inadequate.

Noise levels in the front seats are lower than average and particularly good in the passenger area which gives the impression of being in a much larger aircraft.

To integrate with larger aircraft at busy airports an initial approach may comfortably be flown at 120 kt, if required, going for a V_{at} (target threshold speed) of 80 kt. Trim changes when flaps and undercarriage are lowered are very slight and handling on the approach is flawless. The elevators retain a lot of power at low airspeeds and during the round-out care must be taken not to overcontrol in pitch. Remember that, and landing a Crusader is as simple as arriving in a Cessna 152 (see Chapter 2).

Capabilities

Standard empty weight for a Cessna Crusader includes airways avionics, an autopilot and a number of features usually regarded as extras by other manufacturers. The example tested had a de-icing system and a number of luxury features which added several hundred pounds. Taking this aircraft as an example, maximum ramp weight is 5,175 lb (2,347 kg) and the useful load 1,630 lb (739 kg). Full tanks, 915 lb (415 kg), will allow a payload of 715 lb (324 kg). Four people of average weight 160 lb (73 kg) and 75 lb (34 kg) of baggage could be flown for a with-reserve 850 nm at 10,000 ft using 72 per cent power to achieve 178 kt, or 940 nm if you climb to 12,000 ft, set 60 per cent power and cruise at a still-respectable 167 kt. Fastest cruising speed is 195 kt but you would have to fly at 24,000 ft and I would want a pressurized aircraft for that exercise.

A full cabin – five people and their baggage averaging 200 lb (91 kg) each flown by a 160-lb (73-kg) pilot – represents a cabin load of 1,160 lb (526 kg) and that would allow 470 lb (213 kg) of fuel to be carried, which is 65 Imp/78 US gal (295 litres). Using 72 per cent power at 10,000 ft, range would then be about 420 nm, and is not to be sneezed at for a little bomber.

Verdict

For its power the Crusader does very well. It has a bigger cabin than any light twin of up to 285 hp per engine and its air miles per gallon are particularly attractive (6.7 nm/

Imp gal, 5.55 nm/US gal or 1.47 nm/litre at high cruising speeds). It is cheaper than most of the opposition and, for its power, strong on payload/range. It can match, and in some cases, better other twins with a 70-hp advantage. And it can land at its maximum take-off weight.

To get the best out of a Crusader you should really cruise at 18,000–22,000 ft but we are back to using oxygen masks. There is talk of a pressurized version and that really will be a show-stopper. As it now exists I would say that the Cessna Crusader is one of the finest light twins on the market.

Facts and figures

Dimensions

Wing span	39 ft
Wing area	189.2 sq ft
Length	30 ft, 5 in
Height	13 ft, 4 in

Weights & loadings

Max ramp	5,175 lb (2,347 kg)
Max take-off	5,150 lb (2,336 kg)
Max landing	5,150 lb (2,336 kg)
Equipped empty	3,545 lb (1,608 kg)
Useful load	1,630 lb (739 kg)
Seating capacity	6
Max baggage	590 lb (268 kg)
Max fuel	127 Imp/153 US gal (579 litres)
Wing loading	27.2 lb/sq ft
Power loading	10.3 lb/hp

Performance

Max speed at 18,000 ft	216 kt
72% power cruise at 10,000 ft	178 kt
60% power cruise at 12,000 ft	167 kt
Range at 72% power (10,000 ft)	850 nm
Range at 60% power (12,000 ft)	940 nm
Rate of climb at sea level Two engines	1,480 ft/min
One engine	220 ft/min
Take-off distance over 50 ft	1,750 ft
Landing distance over 50 ft	1,450 ft
Service ceiling Two engines (max operating)	25,000 ft
One engine	13,000 ft

Engines

2 × Continental TSIO-520-AE developing 250 hp at 2,400 rpm, driving 74-in (188-cm) diameter McCauley three-blade propellers.

TBO

2,000 hours.

Piper Aerostar 602P

The high-speed, distance-disposal machine

Background

The Douglas Boston was one of the outstanding American light bombers of the Second World War. After the war Ted Smith, its designer, set up shop and built a series of high-wing light twins under the name of Aero Commander. There was a promising start, then the enterprise hit troubled waters. First an outfit called American Cement bought it (concrete aircraft! what next?) then it was sold to Butler Aviation International. A series of legal actions, out of which only the attorneys made any money, had a happy ending with Ted Smith at the controls of a new aviation concern which traded under his name. By now he had ditched high-wing designs in favour of mid-wing layouts, of which there are very few examples. In fact most mid-wings have been military aircraft. There was a twin jet which eventually became the Israel Aircraft Industries Westwind and a piston-powered streak of greased lightning called the Aerostar. Aerostars come in three versions, the 600 model with normally aspirated Lycoming engines of 290 hp each, the 601 turbocharged Aerostar and the 601P, which is pressurized.

I flew a 601P early in 1976 and apart from a few eccentricities (an anything but foolproof fuel system, a door that needed real muscle to open and close and a zero fuel restriction that damaged its full potential), it was one of the most remarkable and exciting aircraft of *any* category to come my way. Then Ted Smith died – a great loss to world aviation – and the company was taken over by Piper. In 1981 they announced a new improved version of the pressurized model. At first it was named the Piper Sequoya, but a small firm had already laid claim to that Red Indian and they settled the punch-up out of court. Now it is called the Aerostar 602P.

Engineering and design features

The Aerostar is one of the most beautiful small twins of all time. The thin wing has a straight leading edge and a lot of taper from the swept-forward trailing edge (just like

Piper Aerostar 602P, one of the fastest and most fuel-efficient light twins on the market.

Very little space is wasted in the Aerostar flight deck.

Ted Smith's Second-World-War Boston). Two fuel tanks in the wings and a third one in the fuselage have a total usable capacity of 138 Imp/165 US gal (625 litres). There are beautifully designed Fowler flaps with hidden tracking (like the Cessnas), and Frise ailerons. The Lycoming engines reside within wide, flat nacelles. They represent the major change from previous models in so far as integral turbochargers have been fitted to the engines by Lycoming, and a low compression ratio coupled with lower rpm are claimed to have reduced noise levels, increased reliability and endowed the following improvements over the previous 601P: 20 per cent increase in two-engine climb; 17 per cent increase in single-engine climb; 35 per cent increase in single-engine service ceiling, and the ability to maintain the 4.25 psi cabin pressure at throttle settings as low as 40 per cent power. Three-bladed propellers are standard fit.

The tail surfaces are unique. There is a swept fin with a curved leading edge, and the tailplane with its separate elevators is sharply swept back. From the outside it is hard to believe that this is a pressurized aircraft. There is a large, wraparound, single-piece windscreen and the four windows on each side of the cabin are quite massive. This

is one of the most perfectly proportioned fuselages of any small aircraft and, like that of the Partenavia P68C (another elegant design – see airtest earlier in this chapter) it gives the impression of being too small for six people. Yet it is actually 4 in (10 cm) wider inside than the Beech Baron (see next airtest). Standards of engineering are above average and most of the airframe is flush riveted.

Cabin and flight deck
There is a two-piece door on the left side of the fuselage, just ahead of the engine. The top portion hinges up and the lower section drops down to provide a convenient step up to the low-slung cabin floor. The wing mainspar passes through the fuselage behind the rear seats so the cabin provides an unobstructed area 46 in (117 cm) wide, 48 in (172 cm) high and 150 in (381 cm) long. The door position has its advantages and disadvantages, but provided the recommended loading order is used there is no hassle while climbing aboard the lugger. Normal seating is in three pairs with the middle two swivelling to allow business

discussion during the journey. When you have settled and strapped in, the Aerostar is fairly comfortable.

The flight deck looks small because it is crammed with equipment. An annunciator with sixteen warning lights is built into the glareshield. Directly below and in descending order are the three fuel gauges, the avionics station box, the weather radar (essential in a high flyer like the 602P) and the radio installation. Additional radio must be banished to an area left of the captain's wheel. This is a pity; I do not like splitting systems and perhaps it would have been better to make more room above the main avionics stack by placing the fuel gauges elsewhere.

Engine starting, ignition, fuel selectors and boost pumps are in two areas outlined with white lines, one for each engine. They are located left and right of the power quadrant. The electric services are handled on a row of rocker switches lined up along the bottom left of the panel and all circuit breakers are on the far right.

The undercarriage lever moves up and down in a slot cut in the instrument panel to the left of centre and there is a flap control balancing it on the right. It is not of the follow-up type so you must return the switch to neutral when the flap indicator reaches the required setting – not a good arrangement. The flight panel can be based on a director system if required and engine instruments are in front of the copilot's seat.

There is a power pedestal in the centre with the usual pairs of throttle, mixture and propeller levers but you will search in vain for the trim wheels because the job is done by an electric switch on the control yoke for the elevators, and a LEFT/RIGHT rocker on the control pedestal for rudder trim. Also on the pedestal are the pressurization controls. You need a powerful magnifying glass to read the numbers and I give Piper nil out of ten for selecting such equipment when other kit is available that does not set out to make life difficult for the pilot.

This is a somewhat cluttered flight deck, partly because upward vision windows set above the flight deck preclude the use of a roof panel that could have taken some of the controls off the main areas. Also, the entrance door makes it impossible to have a sidewall panel. Piper should take another look at the 602P flight deck – I am sure it could be improved.

In the air
The engines are started in typical fuel-injection fashion: prime with the mixture in RICH, move the levers to CUT-OFF, open the throttle slightly, use the electric starter and move the mixture lever back into RICH when the engine fires. These particular Lycomings object to being idled below 1,000 rpm so the brakes must be used while taxiing to stop the machinery bursting into a gallop. The nosewheel is not connected to the rudder pedals. It is electrically powered and you steer on a little LEFT/RIGHT rocker switch conveniently located on the control console. The switch is spring loaded to the central position and little practice is required to maintain a straight line. V_r is 84 kt,

blue-line speed 117 kt, and a convenient cruise/climb is achieved at 140 kt.

Take-off was quick but vocal and we were settled at 140 kt in no time at all. Ten minutes after lift-off the aircraft shot through 15,000 ft. I was interested to see that a characteristic of the Aerostar's airfoil is that a 1,500 ft/min climb rate can be maintained at reduced power over a speed range of 130–165 kt. Maximum rate at optimum speed (118 kt) is 1,755 ft/min, but at lighter weights she goes up like a rocket at 2,350 ft/min.

Settled at 16,000 ft, 75 per cent power (2,100 rpm/33.6 in manifold pressure) gave an indicated 180 kt which trued out at 234 kt. Noise level under these conditions is about average. At 25,000 ft this power setting will result in a cracking 245 kt, but a more economical and certainly quieter technique is to adopt 65 per cent power, which entails selecting 2,000 rpm and 30 in manifold pressure. At 16,000 ft that will give you 214 kt or, even better, 226 kt at 25,000 ft.

Clean stall came at 90 kt with a slight tendency for the left wing to drop. Maximum flap (40 degrees) reduced that speed by a useful – and I would have thought essential – 20 kt no less. But at the 'g' break the left wing went down to about 80 degrees before recovery action stopped what looked like becoming interesting, particularly for my passengers – the editor of a Belgian aviation magazine and Piper's young demonstration pilot (who I suspect from the way he gripped his seat had never done a full stall in these aircraft before). Possibly the flaps on this particular aircraft were incorrectly balanced, because when I flew an earlier model some years ago it only dropped a wing when stalled flaps-up.

Visibility while turning is excellent and the aircraft handles beautifully. The electric elevator trim is first rate but I thought the rudder trim, operated on the little rocker switch, was inclined to be twitchy. There is slight lateral stability, powerful yaw stability and 1½ cycles were needed to regain a trimmed 180-kt IAS following a 15-kt displacement. Careful use of the trimmers is essential if you are to enjoy the best this remarkable twin-engined projectile has to offer.

To descend quickly at 2,500 ft/min you can lower the wheels and flaps, then depress the nose to maintain 140 kt. The pressurization coped well with this situation and I do not recall any ear-popping. When you shut down an engine to simulate power failure, direction can be maintained on the ailerons alone. At 6,000 ft, loaded at medium weight, I recorded a single-engine climb rate of over 500 ft/min and Piper claim 302 ft/min at maximum weight, which is outstanding for this class of aircraft.

Initial approach may be flown at 120 kt, going for 92 kt over the threshold. Behaviour during this phase of flight is lily-white. The actual arrival is simplicity itself and not a lot faster than usual for a medium-powered light twin. As the speed declines and the rudder becomes less effective, direction must be maintained on the electric nosewheel steering.

Capabilities

Maximum ramp weight is 6,029 lb (2,735 kg) and a well-equipped Aerostar 602P would support a useful load of 1,779 lb (807 kg). Assuming full tanks, 993 lb (450 kg), payload comes out at 786 lb (357 kg). That could be used to carry a pilot and three passengers along with more than 100 lb (45 kg) of baggage, assuming each person weighs an average of 170 lb (77 kg). Alternatively, five people of average weight 157 lb (71 kg) could be flown without

baggage. With full tanks the with-reserve range is 1,070 nm at 65 per cent power, cruising above most of the weather at 23,000 ft, while cracking along at 222 kt. With six 170-lb (77-kg) occupants and 240 lb (109 kg) of baggage (that is, a full cabin) 519 lb (235 kg) remains for fuel, which translates into 72 Imp/86 US gal (326 litres). And that would allow a still respectable range of 550 nm. So in terms of payload/range the 602P does better than the average twin.

Verdict

The Piper Aerostar 602P represents a first step into the high-performance arena. It is small, heavy, fast but capable. I would like it more if Piper had simplified such details as electric fuel cocks (do we really need that in a little aircraft?) and electric nosewheel steering (same question). Although Piper have reduced the noise level slightly it is still not a quiet aircraft. On the plus side, there is no landing weight restriction.

Being a fast, relatively complex animal with a pressurized cabin, the Aerostar 602P cannot be cheap. But in terms of high-performance economy it beats all comers. It is the only plane capable of cruising six people at over 220 kt for a fuel burn of 180 lb (82 kg)/hour.

With six people the Aerostar 602P cruises at World War Two fighter speeds on a fraction of the power.

Facts and figures

Dimensions

Wing span	36 ft, 8 in
Wing area	178 sq ft
Length	34 ft, 9 in
Height	12 ft, 1 in

Weights & loadings

Max ramp	6,029 lb (2,735 kg)
Max take-off	6,000 lb (2,722 kg)
Max landing	6,000 lb (2,722 kg)
Equipped empty	4,250 lb (1,928 kg)
Useful load	1,779 lb (807 kg)
Seating capacity	6
Max baggage	240 lb (109 kg)
Max fuel	137 Imp/165 US gal (625 litres)
Wing loading	33.7 lb sq ft
Power loading	10.3 lb/hp

Performance

Max speed	262 kt
75% power cruise at 25,000 ft	245 kt
65% power cruise at 23,000 ft	222 kt
Range at 65% power (23,000 ft)	1,070 nm
Range at 55% power (202-kt cruise at 23,000 ft)	1,150 nm
Rate of climb at sea level Two engines	1,755 ft/min
One engine	302 ft/min
Take-off distance over 50 ft	2,250 ft
Landing distance over 50 ft	2,076 ft
Service ceiling Two engines	28,000 ft
One engine	12,900 ft

Engines

2 × Lycoming IO-540-AA15A developing 290 hp at 2,425 rpm, driving Hartzell, 78-in (198-cm) diameter, three-blade propellers.

TBO

1,800 hours.

Beechcraft Baron 58P

A pressurized Bonanza with two engines

Background

Beechcraft have rightly collected a reputation for high-quality if somewhat conventional engineering. Certainly their aircraft are built to last and, in general aviation terms, to work hard (commercial aircraft must withstand far higher utilizations to earn their keep).

At one time Beech, as they are often known, produced a development of their outstanding single (see Chapter 4), which they called the Twin Bonanza. It was quite a large aircraft with a wide cabin giving 12 in (30 cm) more elbow room than a Bonanza and turning the scales at 7,300 lb (3,311 kg). They offered it with engines of 295 or 340 hp but with a wing span of almost 46 ft. It was certainly a big bird for a six-seater.

The subject of this airtest has a stronger claim to the title Twin Bonanza because its airframe is derived largely from the Bonanza A36. However, I have long since given up trying to fathom the model number and naming policy

of aircraft manufacturers and for reasons best known to Beech they have titled what is truly a twin-engined Bonanza the Beechcraft Baron. Several versions are on offer. The model featured here is their top-of-the-range, pressurized Baron 58P.

Engineering and design features

Standard fuel capacity is 138 Imp/166 US gal (627 litres) but there is an optional 158-Imp/190-US-gal (719-litre) system. All fuel is carried in the wings. The slotted flaps, undercarriage and engine cowl flaps (for controlling engine temperature) are all worked electrically. The undercarriage takes only 4½ seconds to retract and it has been drop-tested at sink rates of up to 600 ft/min. Continental turbocharged engines, each of 310 hp, hide within low, flat-profile cowlings, and three-blade propellers are fitted as standard. The exhaust system is made of stainless steel; surely the only suitable metal for planes or cars.

There is a very large swept fin with a dorsal area extending forward to the passenger cabin and the rudder has a trim tab of greater than usual area. Beech have remained faithful to fixed tailplanes and separate elevators. The fuselage is basically that of the Bonanza A36 with the exception that the pressurized Baron does not have the large double doors on the right-hand side. Instead there is a single one on the left giving access to the four cabin seats which may be arranged club fashion in facing pairs. The separate door on the right for the two front seats remains unchanged. Although the 58P is pressurized to 3.7 psi, Beech have managed to retain their large windows (four on each side) and one-piece windscreen.

A baggage compartment in the nose, with its door on the right, has a capacity of 300 lb (136 kg) but another 400 lb (181 kg) may be carried at the rear of the cabin with yet another 120 lb (54 kg) in an 'extended rear compartment', making a potential total of 820 lb (371 kg) if you want to move house. (Can you imagine six people each wanting to take 136 lb (62 kg) of gladrags, golf clubs and corkscrews? Some party that would be.)

Although the airframe is peppered with snap-head rivets, a quick walk-around soon reveals that the Baron is no ordinary light twin. Standards of engineering and finish are several cuts above average.

Cabin and flight deck

Being basically a Bonanza fuselage, the Baron 58P suffers from a 42-in (107-cm) wide cabin, which is 4 in (10 cm) narrower than a Piper Aerostar, 6 in (15 cm) tighter than the Cessna Crusader, while giving away 7 in (18 cm) to the Piper Seneca (see respective airtests in this chapter). To some extent Beech restore honour by providing another few inches' headroom, a lot of leg-room, the deficiency in width being partly redeemed by the door arrangement which dispenses with the need to provide an aisle between

OPPOSITE: Beech Baron 58P, a twin-engine development of Beechcraft's very successful Bonanza.

the four passenger seats. If you recall, I did mention that the rear door at the left opens up on club seating so you can step in, turn right to sit facing ahead or left if you want to look backwards (and see where you have been). As an option you can have a small table that extends from the right wall, opposite the door, and positions itself between the two facing pairs of seats.

Overhead are individual reading lights and fresh-air vents. Generally the cabin area is reasonably comfortable.

The flight deck is very similar to, and just as old-fashioned as, that of the single-engine Bonanza. For example, standard aircraft have a throw-over control wheel which operates from an arm pivoting on a tube that emerges from the centre of the instrument panel. When you want full dual control, and many folk do when they buy this class of aircraft, they attach a massive beam to the central tube and stick a control yoke on each end – a crude arrangement that obscures some of the instruments.

The flight instruments are on a separately sprung panel with a bank of switches below them for the electrics, all of them good-quality but old-fashioned tumblers. Immediately above the tube that carries the control-yoke beam is a rather high-set power quadrant. For some reason Beech arrange the pairs of levers in a rather unusual fashion. On the left are the propeller levers, in the centre are the throttles and to the right are the mixture controls. I am not keen on this rather nonstandard layout and found the levers a little too high for comfort.

Above the engine controls are the related instruments, including fuel gauges which read ¼, ½, ¾ and FULL. I would like to see these calibrated in pounds. Radio is in a stack to the right of centre and further right again are some engine temperature and pressure gauges. Extending below the power quadrant and the main control tube (which moves in and out as you push or pull the control yokes) is a small console which carries the elevator, rudder and aileron trim wheels, the cabin pressure controls and the fuel selectors. The flight deck reeks of quality, but planning is a little untidy and it reminds me of the 1930s

In the air

After start-up it was interesting to note the drop in noise level as soon as the pressurization was switched on and the door seals had inflated. Nosewheel steering is excellent and so are the brakes. The traditionally American high-set instrument panel tends to obscure the view ahead to some extent, otherwise visibility on the ground is good for a pressurized aircraft.

On the day of my flight there was a 25-kt wind blowing 20 degrees to the right of the runway but this did not seem to annoy the Baron, so I was happy too. V_{mca} is 80 kt and we were soon past that figure, heading for the 110-kt blue-line speed. There is no shortage of steam during take-off; 38 in of manifold pressure see to that.

During a cruise/climb at reduced power and 130 kt I timed 1,900 ft/min and this was maintained up to 10,000 ft. Maximum climb rate for a fully loaded 58P is 1,529 ft/

min. Like all aircraft capable of flying within a wide band of flight levels, cruising speed and range performance varies according to altitude and the power setting adopted.

Although no percentage power figures are quoted in the Beech Baron manual, their sales literature talks of 81, 74, 65 and 56 per cent and gives cruising speeds – which vary from 186 to 241 kt – against altitudes of 15,000, 20,000 and 25,000 ft. Noise levels are low enough for people in the club area to overhear what the pilot and his neighbour are saying (which may or may not be a good thing!). Handling is outstanding in every respect, and the Baron feels what it is: a Rolls-Royce among light twins. Stability is of a similarly high order, with slow, stately returns to a trimmed speed after the nose has been lifted and the wheel released. Visibility while in the air is excellent except for straight ahead, when the instrument panel tends to intrude. With maximum flap (30 degrees) the aircraft stalls at 77 kt, which is 7 kt slower than the clean stall. There was a slight tendency for the right wing to drop on the example tested.

Like its little sister, the Duchess (see airtest in this chapter), you can close a throttle on the 58P and check the yaw with aileron while the feet are off the rudder pedals and that is proof of good aileron design. Well done Leslie Frise – what would we have done without you? The rudder trim will return the aircraft to balanced asymmetric flight and there are no problems while flying on one engine except for the usual one that plagues most light twins – poor single-engine climb performance. The Baron 58P goes up at 204 ft/min when fully loaded and that, as I said at the start of this chapter, is not good enough.

Although the maximum cabin pressure differential is only 3.7 psi, it is sufficient to maintain a sea-level environ-

The Baron looks good in the air.

The somewhat dated flight deck suffers from a clumsy dual control arrangement and this in turn has made it necessary to locate the engine controls in a high and not particularly comfortable position.

ment up to an aircraft altitude of almost 8,000 ft. At 21,000 ft cabin altitude is 10,000 ft and a warning light tells you to use the oxygen masks if the cabin climbs any higher. This could be caused by climbing above the limits of the pressurization system or loss of cabin pressure when an emergency descent is made by lowering the undercarriage, selecting flap 15 degrees and stuffing down the nose until 175 kt is showing on the ASI. Using that technique you can descend from 25,000 ft to 10,000 ft in about 10 minutes or less. The pressurization system works well and at no time was I aware of ear discomfort.

While I had been enjoying myself cloud dodging in this beautiful aircraft the wind had taken a turn for the nasty back at our single-runway destination with gusts reported at up to 40 kt. Under more normal conditions an initial approach at 120 kt followed by 100 kt over the threshold would be ideal. Because of the gusting wind I was advised to use 110 kt for short finals and this resulted in a rather lengthy float, despite the wind. All controls remain effective during the final stages of landing and even on the miserable day of my flight the Baron 58P proved easy to land.

Capabilities

Maximum ramp weight is 6,140 lb (2,785 kg) and the empty weight of a well-equipped 58P (full airways radio, autopilot, de-icing, club interior, etc) would on average be 4,240 lb (1,923 kg), leaving a useful load of 1,900 lb (862 kg). Fill the tanks, 1,140 lb (517 kg), and you are left with a payload of 760 lb (345 kg), which is enough for a pilot and three people with 120 lb (54 kg) of luggage. The aircraft could fly such a load for 1,030 nm at 20,000 ft using 74 per cent power to produce a cruising speed of 223 kt, or you can add 100 nm to the range if 65 per cent power is adopted to attain 210 kt at the same altitude.

Six people of average 170-lb (77-kg) weight and 240 lb (109 kg) of baggage would represent a payload of 1,260 lb (572 kg) and allow 640 lb (290 kg) for fuel, ie, 88 Imp/ 106 US gal (401 litres), when at the altitudes and power settings already mentioned the ranges become 552 nm at 223 kt and 626 nm at 210 kt. The Aerostar 602P will carry the same load at 65 per cent power 12 kt faster over a range of 550 nm. There is not a lot of difference in field performance between the two aircraft.

Verdict

When you move into the pressurized twin class, the aircraft have got to be good. The nearest competitor of the Baron 58P is the similarly powered Cessna 340 which offers a slightly wider fuselage (about the same as the Aerostar), the charms of a cabin class aircraft with let-down airstairs, a centre aisle and a more modern flight deck. In terms of fuel economy expressed as air miles per gallon the most effective is the Aerostar 602P, then the Baron 58P, followed by the Cessna 340 which at 75 per cent power and 20,000 ft is 7 kt slower than the Baron, giving away 24 kt to the Aerostar. When comparing engineering quality

the Baron comes out best, but so it should – it is the dearest in its class. And by the way, you can land the brick-built Baron at its maximum take-off weight.

Like all purchasing decisions other factors than numbers on a sheet of paper must affect the final choice. The Baron may have a narrow cabin but there is plenty of leg-room and height. It is also very quiet. The Aerostar might be vocal but its fuel economy is unmatched. The Cessna 340 is a good all-rounder, cheaper than the others and with a lot of charm. Take your pick!

Facts and figures

Dimensions

Wing span	37 ft, 10 in
Wing area	188.1 sq ft
Length	29 ft, 10 in
Height	9 ft, 6 in

Weights & loadings

Max ramp	6,140 lb (2,785 kg)
Max take-off	6,100 lb (2,767 kg)
Max landing	6,100 lb (2,767 kg)
Equipped empty	4,240 lb (1,923 kg)
Useful load	1,900 lb (862 kg)
Seating capacity	6
Max baggage	420 lb (191 kg)
Max fuel	
Standard	138 Imp/166 US gal (628 litres)
Optional (as tested)	158 Imp/190 US gal (719 litres)
Wing loading	32.4 lb/sq ft
Power loading	9.8 lb/hp

Performance

Max speed	255 kt
74% power cruise at 20,000 ft	223 kt
65% power cruise at 20,000 ft	210 kt
Range at 74% power	1,030 nm
Range at 65% power	1,130 nm
Rate of climb at sea level	
Two engines	1,529 ft/min
One engine	204 ft/min
Take-off distance over 50 ft	2,376 ft
Landing distance over 50 ft	2,498 ft
Service ceiling	
Two engines	above 24,000 ft
One engine	14,400 ft

Engines
2 × Continental TS10-520L developing 310 hp at 2,700 rpm, driving 78-in (198-cm) diameter, three-blade propellers.

TBO
1,600 hours.

Piper PA-31 Navajo

Best-selling cabin-class light twin

Background
One of the first, if not *the* first of the cabin-class light twins to appear on the general aviation scene was the Piper Navajo. It came onto the market in 1960, took off with a vengeance and has been a shining light ever since. Piper have good reason to be proud of their PA-31 Navajo. In the first place it outsells competing cabin-class twins, and in addition the basic airframe has led to the development of Piper's turboprop family: the Cheyenne I, II (see Chapter 6) and III, not to mention a few that have collected such odd names as the Cheyenne IIXL.

In years to come the Navajo will be regarded as one of the classics of general aviation. Not that it is as pretty as some of the Cessna light twins. But the big feature of the aircraft is its intensely practical concept, a characteristic which time and again has proved one of the biggest plus marks in any buyer's market.

Engineering and design features
The wing contains four flexible fuel cells, two on each side, with a total usable capacity of 156 Imp/187 US gal (708 litres). Under normal conditions each engine is supplied by its adjacent tanks but during asymmetric flight the usual crossfeed facility is available.

The electrically operated flaps have a most complicated system of rheostats, power solenoids and a small amplifier to ensure that both sides come down together – what a performance! When I think of the brass lever and hand pump fitted to the old Ansons I flew on occasions during the Second World War – the flaps were guaranteed to come down one at a time but we managed without all that black magic fitted to the Navajo. Surely Mr Piper there must be an easier way to keep your flaps in step.

Two turbocharged Lycoming engines are fitted, each developing 310 hp. They drive three-blade propellers and each engine has a 28-volt, 70-amp alternator to ensure adequate electric supply.

For its size the Navajo has the biggest tailplane of all time – half the span of the wings – and it carries equally generous elevators. The swept fin extends along the roof, blending nicely into the non-tapered section of the fuselage. Five windows, four of them very large, are let into the fuselage sides and I was intrigued during a visit to

The Piper Navajo, biggest seller among the cabin-class light twins, showing the large, two-piece entrance door.

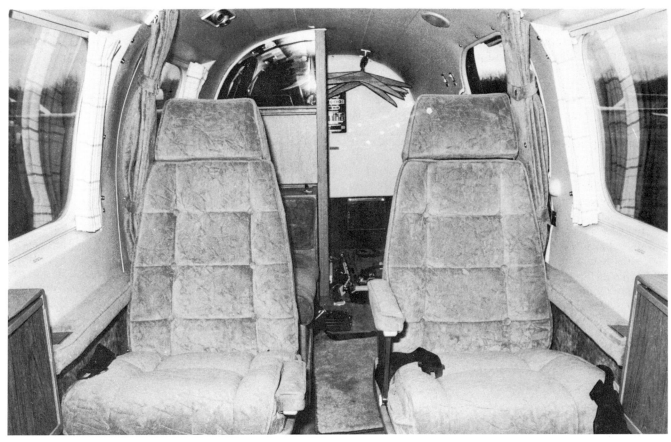

The comfortable cabin with its big windows is a bright and cheerful environment in which to travel.

Piper's Lakeland Florida factory to see that they first completely clad the fuselage before cutting out the window openings with a posh can opener – and why not? At the rear of the cabin area, on the left-hand side, is a two-piece entrance door, the top portion hinging up and allowing the lower part to drop down and become a set of airstairs. There is a 150-lb (68-kg) baggage locker in the nose and another 200 lb (91 kg) may be carried in the luggage area at the back of the cabin. Optional lockers built into the rear of the engine nacelles offer an additional 300 lb (136 kg) of baggage capacity.

The Navajo fuselage is not particularly long for an aircraft in its category (hence the massive tailplane), yet it provides a roomy cabin. So on the ground the aircraft stands high and looks a little chunky, but not unattractive.

Cabin and flight deck
Entry and exit through the very practical door with its built-in steps could not be easier. The cabin is 50 in (127 cm) wide and 51½ in (131 cm) high – not exactly in the 'stand-up' class but large by light-twin standards. The four seats in the passenger area of the example flown were arranged club fashion (in facing pairs) with a quite generous centre aisle, which makes it easy to enter or leave the aircraft and confers an air of spaciousness to the cabin. This is further enhanced by the exceptionally large windows which make the interior a light and cheerful

place. Pull-up and fold-out tables are provided for each facing pair of passengers and there are the usual passenger amenity panels adjacent to each seat. These provide a reading light, an adjustable fresh-air vent and a built-in oxygen point for occasions when it is planned to make use of the aircraft's ability to cruise high. Then it would be necessary to wear oxygen masks because the Navajo is not pressurized. The seats are better than anything you will find on an airliner, having large headrests, soft padded arms and splendid velour covers.

Talking of seats, those in the flight deck area are adjustable in most directions. The circuit breakers are confined to the left side wall and so is the battery master switch but the manufacturers have rather illogically banished the ammeter and the alternator test buttons to the roof panel where they reside between a pair of fuel gauges. In an ideal flight deck all related controls should be grouped together.

Ignition switches, electric fuel pump switches, the various electric services, de-icing controls and a LEFT/ RIGHT rocker switch for starting the two engines are all on the roof panel. The fuel selectors are on the floor between the two front seats, and although it all looks neat and tidy I suspect the layout has been determined by engineers, not pilots, because of the way related controls have been spread around the place. Likewise the engine readouts are split: rpm, manifold pressure, fuel flow, and exhaust gas temperature run along the top of the centre panel, while the temperature and pressure gauges, most of them 'two-in-one' instruments giving readings for both engines, are on the bottom right-hand panel. This arrangement does however leave a very large area in the centre for avionics where they can easily be reached by either pilot.

Under the glareshield is a line of annunciator lights. A wheel-shaped lever positioned to the right of the captain's control yoke works the undercarriage and the flap control is on the other side of the central pedestal. Piper have managed to keep their wide and deep instrument panel below eye-level, showing that it can be done.

The pedestal carries pairs of throttles, mixture controls and propeller levers, trimmers (and their position indicators) for the elevators, ailerons and rudder, the autopilot control panel and, at the bottom of the stack, two engine-cowl flaps-control switches with their position indicators.

While I feel that Piper might have done a better planning job on the flight deck, it is a pleasing and easy-to-live-with working area, and, like all things in life, you can get used to finding your way around.

In the air

Unfortunately, circumstances on the day of my airtest prevented our arranging for a representative load, so this was a rather light Navajo. Nosewheel steering is very precise but rather heavy and the brakes are excellent by any standards. For the take-off, V_r was 76 kt going for 90 kt after lift-off. Blue-line speed that would provide best rate of climb if an engine failed is 94 kt. Application of power produced a lusty sprint down the runway. We lifted off and at 90 kt I raised the wheels and flaps, 15 degrees having been selected for take-off. Indicated rate of climb was 1,600 ft/min and this we maintained up to 10,000 ft. At 10,000 ft, 2,300 rpm/33 in manifold pressure (representing 75 per cent power) gave a TAS of 188 kt using economy mixture, and 65 per cent produced 175 kt with a much quieter cabin. Although the Navajo is not a noisy aircraft I do not believe it is as quiet as some of the Cessna twins of similar power. Visibility is above average in all modes of flight.

The Navajo is a satisfying aircraft to fly with nicely harmonized controls, my only complaint being the rudder trim, which is too low geared. When later in the flight I shut down one engine it was just about able to trim out all rudder loads. Lateral stability is neutral (roll on bank, let go of the wheel and it will fly with one wing down all day). Directionally it behaves like an arrow and 1½ cycles were taken to regain a trimmed 170 kt IAS following a 15-kt artificially-induced upset.

At maximum weight the Navajo will climb on one engine at 245 ft/min – not as bad as some planes in this chapter but still not good enough. Of course, like all aircraft, the engine failure after take-off situation becomes

progressively less critical at lower operating weights. The first 15 degrees of flap may be lowered at 174 kt and full-flap (40 degrees) can come down at up to 140 kt. Trim changes are slight and full-flap reduces the stalling speed from 72 to 62 kt.

Initial approach was flown at 120 kt with 90 kt over the threshold. It is easy to fly accurately and it is no more demanding to land than a light single.

Capabilities

For its power the Navajo is quite a big bird with a maximum ramp weight of 6,536 lb (2,965 kg). A very well-equipped example would have an empty weight of 4,350 lb (1,973 kg), so the useful load for such an aircraft would be 2,186 lb (992 kg). Full tanks weigh 1,123 lb (509 kg), leaving a payload of 1,063 lb (482 kg), which is enough for a 170-lb (77-kg) pilot and five passengers of the same average weight. Between them they could bring 43 lb (20 kg) of baggage; but average weights are usually less than 170 lb (77 kg) a head, and the baggage allowance can increase accordingly. With full tanks you have a with-reserve range of 995 nm at 196 kt (adopting 75 per cent power at 12,000 ft) or 1,050 nm when the 65 per cent power setting is used to attain a cruising speed of 185 kt.

When it is necessary to carry a pilot and five passengers plus the usual airline allowance of 44 lb (20 kg) luggage for the customers cabin load becomes 1,240 lb (562 kg), leaving 946 lb (429 kg) for fuel, 131 Imp/158 US gal (598 litres), and the respective ranges for 75 and 65 per cent power become, in round figures, 820 and 860 nm. It is when you start going into the numbers in this way that the reason for the Navajo's success becomes clear. Among other attractive features, its above-average payload/range makes it a highly practical light transport aircraft.

Verdict

As I said before, the Navajo does not have the grace of comparable Cessna twins, the nearest rival being their 402 Businessliner, which is slightly more powerful and heavier. However, the strength of any design lies in its ability to do a job of work as cheaply and efficiently as possible. In this respect the Piper Navajo scores high marks. It is a good all-rounder. The Navajo might not be the fastest 310-hp twin flying but it is by no means slow. It has a good field performance, above average air miles per gallon (about 8.4 nm per Imperial and 7 nm per US gallon, or 1.85 nm per litre, at 75 per cent power when cruised at 16,000 ft). It can also be landed at its maximum take-off weight, which is a distinct operational advantage.

Most of the engineering problems of earlier models (and there were a few) have now been overcome and Navajos are in worldwide service as corporate aircraft, commuter planes, charter planes and advanced trainers for airline pilots. Some even spend their working lives flying vacationers on sightseeing trips down the Grand Canyon.

My only real criticism of the Piper Navajo is its not very inspiring single-engine climb. When flown at 6,000 lb

Although a generous cabin in a small aircraft is bound to make for a chunky appearance, the Navajo looks attractive in flight.

(2,722 kg) there is an acceptable 350 ft/min, but the aircraft has a maximum take-off weight of 6,500 lb (2,948 kg). And that extra 500 lb (227 kg) takes more than 100 ft/min off the engine-out climb rate. One or two of the firms specializing in extracting more performance from existing designs have come up with a 'fix' for the Navajo's single-engine climb. If they can do it, why not Piper? Their PA-31 is an outstanding aircraft in most respects and it deserves a single-engine climb to match. Having said this, there can be no doubt that the Piper Navajo has earned its distinction of being the best-seller in its class.

Facts and figures

Dimensions

Wing span	40 ft, 8 in
Wing area	229 sq ft
Length	32 ft, 7 in
Height	13 ft

Weights & loadings

Max ramp	6,536 lb (2,965 kg)
Max take-off	6,500 lb (2,948 kg)
Max landing	6,500 lb (2,948 kg)
Equipped empty	4,350 lb (1,973 kg)
Useful load	2,186 lb (992 kg)
Seating capacity	6
Max baggage (in fuselage)	350 lb (159 kg)
Max fuel	156 Imp/187 US gal (708 litres)
Wing loading	28.4 lb/sq ft
Power loading	10.5 lb/hp

Performance

Max speed	227 kt
75% power cruise at 12,000 ft	196 kt
65% power cruise at 12,000 ft	185 kt
Range at 75% power (12,000 ft)	995 nm
Range at 65% power (12,000 ft)	1,050 nm
Rate of climb at sea level Two engines	1,220 ft/min
One engine	245 ft/min
Take-off distance over 50 ft	2,095 ft
Landing distance over 50 ft	1,818 ft
Service ceiling Two engines	24,000 ft
One engine	15,200 ft

Engines
2 × Lycoming TIO-540-A2C developing 310 hp at 2,575 rpm, driving Hartzell 80-in (203-cm) diameter, three-blade propellers.

TBO
1,800 hours.

Cessna 404 Titan

A giant among light twins

Background

It stands among the business and privately owned hardware, towering over most else on the parking ramp. Yet the big and beefy Titan has the nature of a lamb. Cessna have a wide range of light twins and while to some extent they do transfer bits of one design to make up another, the practice is not pushed to the same limits as at Piper.

Top of Cessna's piston-engine range is their high-flying, high-performance Golden Eagle, which is featured in the next test report. By adding some 3 ft to the Golden Eagle's fuselage and 5 ft to its span they have evolved the Titan. It is an unpressurized aircraft that is offered in two versions: the Titan Ambassador, fitted out as a business aircraft; and the Titan Courier, a light freighter which can quickly be converted to a commuter plane with up to eleven seats in its commodious cabin. This airtest relates to the Titan Courier.

Engineering and design features

The wing is a bonded structure, almost devoid of rivets, with a surface finish like glass. Frise ailerons are accompanied by large-area Fowler flaps – both devices being the best of their type of control available to the aircraft manufacturer. This is a wet wing with a usable capacity of 284 Imp/340 US gal (1,287 litres). A pair of 375-hp Continental engines is installed wide apart so that the propeller tips are kept well away from the fuselage. Close-fitting cowlings are designed to separate hot and relatively cold running sections of the engines. The advantage of keeping propeller and slipstream noise as far away from the cabin as possible is self-evident. The disadvantage is that during asymmetric flight an engine situated further than usual from the aircraft's centre-line will create a more powerful yawing moment which must be contained by providing a balancing rudder force. A large fin and rudder has managed to hold V_{mca} to a not particularly frightening 78 kt, which is only 2 kt faster than a Navajo (see previous report). The tailplane and its elevators have pronounced dihedral angle.

The lever-suspension undercarriage is raised and lowered by engine-driven hydraulic pumps with a back-up emergency nitrogen bottle in the event of both pumps going on the blink. The flaps are also powered off this system. Baggage may be carried in the carpeted areas provided at the back of each engine nacelle, 200 lb (91 kg) in each, 350 lb (159 kg) in the nose bay with up to another 250 lb (113 kg) in the adjacent avionics bay (according to the weight of radio equipment installed therein). Then another 500 lb (227 kg) may be placed in two bays aft of the rear seats so, in all, there is a potential baggage capacity of 1,500 lb (681 kg) which, of course, is utter nonsense in an aircraft of this class, even a Titan. 500 lb (227 kg) is more than enough for a ten-passenger light twin. I really do not understand why some manufacturers make such a big deal of quoting baggage capacities that bear no relationship to the real world and which, if used with all seats occupied, would allow enough fuel for a trip to the next airfield, if it were not too far away.

The fuselage is long, sleek and full of windows, seven on each side, with a double door behind the wing on the left which opens to reveal a cargo entry/exit point measuring 49 in (124 cm) by more than 50 in (127 cm) high – some hole for a light twin. During passenger-only flights the left-hand door remains closed and the right portion opens in two halves, top section upwards and the lower part down towards the ground. Two steps are built-in and there is a flexible hand rail to assist the fare-paying customers. So that the pilot(s) may enter or leave the aircraft when it has been filled from floor to ceiling with freight, the captain's left-hand side window opens up and outwards. He is then supposed to climb onto the wing root (a nonslip walkway is provided) – not a good arrangement, particularly for a slightly well-fed airframe driver.

The Cessna Titan, with its big three-blade propellers, is a good-looking aircraft on the ground. In light aviation terms it is of impressive size.

Cabin and flight deck

With a cross-section 56 in (142 cm) wide and 50 in (127 cm) high there is no shortage of room. Spaciousness is emphasized by the specially designed seats with their high backs and headrests. There are no arms in the Courier seats and this allows a centre aisle of more than 11 in (30 cm). Five pairs of seats make the Titan Courier a useful commuter plane. The cabin is big, bright and cheerful and, if required, another seat may be added making this a pilot and ten-passenger light transport aircraft.

Up front in the working-class end of the ship is a beautifully designed flight deck. A large central pedestal carries the engine controls, trimmers and autopilot panel. Circuit breakers are on the right sidewall and the left ledge below the captain's window. Here he can easily reach the magneto, starting and priming switches, emergency alternator field switches and most of the electric services. There is a combined voltmeter/ammeter and all related circuit breakers are to hand, those on the right sidewall panel being confined to the radio and autopilot.

Avionics are carried in three vertical stacks in the centre

of the main panel; all the engine instruments are in a line along the top of the panel; and there is room for two sets of flight/radio navigation instruments, one for each pilot.

On the far left is a twenty-two-light annunciator, the undercarriage switch is to the left of the power console and on its right is the flap lever, a clever device which works as follows. Select TO and APPR (take-off and approach) and the inboard flaps roll out then depress to 10 degrees while the outboard sections go down 8 degrees. In the LAND position the flaps extend fully with the inner sections depressed to 35 degrees and the outers down 23 degrees. Move the flap selector to UP and all sections move back and hide themselves within the wings. Fuel is managed on a pair of selectors which may be set to ON, OFF and CROSSFEED. This magnificent flight deck is crowned by a pair of very comfortable pilots' seats that have push-button adjustment of back angle, height and leg reach. These modern flight decks have come a long way since I was a flying instructor on Tiger Moths.

In the air
The engines fire up easily, grumbling into life and idling with a sound that reminds one of the piston-engine airliners that used to grace the airports of the world. Taxiing could not be simpler and nosewheel steering is light for an aircraft weighing more than 8,000 lb (3,629 kg). V_{mca} is 78 kt and blue line is marked at 109 kt. I opened the throttles against the brakes, allowing the turbochargers to stabilize at 28 in, then released the brakes before firewalling the throttles. A waste gate ensures that

40-in manifold pressure is not exceeded. At 90 kt a gentle backward movement of the control yoke lifted us cleanly off the ground. Best rate-of-climb speed is 108 kt and our lightly loaded Titan went up at 2,300 ft/min. A fully loaded one will climb at a still sprightly 1,575 ft/min.

Maximum cruise power recommended by Cessna is 77.5 per cent when a Titan at maximum weight should do you 196 kt at 10,000 ft or 213 kt if you are prepared to suffer oxygen at 20,000 ft. In any case a Titan Courier flown on passenger routes would spend most of its time cruising at 8,000–12,000 ft or lower. For example, at 2,500 feet, 1,800 rpm/33.5 in manifold pressure (75 per cent power) gave an IAS of 185 kt (TAS 190), although that figure was about 5 kt faster than book value because of our very light weight at the time. Noise levels during the take-off, climb and cruise are lower than average, largely due to the geared engines which drive the propellers at relatively low speeds in the range of 1,600–1,900 rpm while cruising, and only 2,235 rpm at maximum power.

The Titan is a pussy-cat during stalling. Flaps-up the 'g' break came at 74 kt and in the landing configuration (maximum flap, wheels down) the ASI read 61 kt. Wings remained almost level flaps-up or down.

All controls, including the trimmers, work beautifully. Visibility is excellent for the flight deck crew (it is pretty good for the customers too) and the Titan is a pleasure to

Cessna Titan, a giant of a light twin.

ABOVE: The Titan is ideal for low-density commuter routes or as a relatively inexpensive company transport.

BELOW: When used as a freighter an adjacent door (left of airstairs) opens to allow stowage of large items.

fly, my one reservation being a complete lack of lateral stability, it being neutral in that plane. I shut down the left engine, and to maintain direction rudder load was modest. The trimmer was powerful enough to remove out-of-balance forces, feet off the pedals. Single-engine climb at maximum weight is, at only 230 ft/min, worse than the Piper Navajo and not good enough.

On the approach there were hardly any trim changes when the wheels and flaps were lowered. At 100 kt the aircraft was rock steady, and following 90 kt over the threshold the round-out was completed, power was brought back to the idle stops, a little back pressure on the still-powerful elevator control got us in the slightly nose-high attitude and we touched down mainwheels first. Any club pilot of average ability could land a Titan.

Capabilities

Strangely, Cessna quote no maximum ramp weight for their Titan, but maximum take-off is 8,400 lb (3,810 kg). Since the aircraft would have to burn off 300 lb (136 kg) of fuel, ie, 41 Imp/50 US gal (189 litres), before landing at the maximum of 8,100 lb (3,674 kg), as a short-distance commuter aircraft you could probably not operate the Titan at more than 8,200 lb (3,720 kg).

Taking its maximum weight as a starting point, an averagely equipped example would have an empty weight of 5,000 lb (2,268 kg), leaving 3,400 lb (1,542 kg) of useful load. Full tanks, 2,040 lb (925 kg), would allow 1,360 lb (617 kg) in the cabin, which is enough for a 170-lb (77-kg) pilot and 6 passengers with the usual airline baggage allowance. It would then fly a with-reserve 1,388 nm, cruising at 196 kt when the 77.5 per cent power setting is used at 10,000 ft, or almost 1,600 nm at 65 per cent power when the cruising speed becomes 181 kt. However, the real talent of a Cessna Titan is to fly 9 passengers economically. A cabin load of 1,970 lb (894 kg), say 9 passengers and their baggage plus the pilot, would allow a fuel uplift of 1,430 lb (649 kg), which translates into 199 Imp/238 US gal (901 litres) and that would be enough to fly a with-reserve 950 nm at 196 kt, or 1,100 nm at 181 kt. In terms of passenger miles per gallon, 43 nm/Imp gal and 36 nm/US gal while cruising at 196 kt has got to be good news; and so has 9.45 passenger nm/litre. Alternatively, a number of short sectors could be flown without having to refuel. As a light freighter the Titan will fly a 3,000-lb (1,361-kg) load for about 180 nm. All this adds up to a very capable aircraft.

Verdict

The Cessna Titan is more powerful and considerably more expensive than the Piper Navajo Chieftain (bigger brother of the Navajo described in the previous test). However, it has a larger cabin, about 700–800 lb (318–363 kg) more useful load and greater range capabilities.

Apart from the Titan's miserable single-engine climb it is a very satisfactory aircraft. However, its two Continental engines are complex machines with reduction gearing to

the propeller drive and they have a lower overhaul life than the Lycoming units in the Chieftain. Much depends on how these engines are treated in service. Expose them to ham-fisted handling and they will give trouble; treat them with consideration and they are known to provide trouble-free hours up to their TBO. But the engines are a little frail and they *must* be handled with care.

The Titan is a fine light-passenger transport or freighter and a pilot's friend into the bargain.

Facts and figures

Dimensions

Wing span	46 ft, 4 in
Wing area	242 sq ft
Length	39 ft, 6 in
Height	13 ft, 3 in

Weights & loadings

Max take-off	8,400 lb (3,810 kg)
Max landing	8,100 lb (3,674 kg)
Max zero fuel	8,100 lb (3,674 kg)
Equipped empty	5,000 lb (2,268 kg)
Useful load	3,400 lb (1,542 kg)
Seating capacity (with crew)	11
Max baggage	1,500 lb (680 kg)
Max fuel	284 Imp/340 US gal (1,287 litres)
Wing loading	34.71 lb/sq ft
Power loading	11.2 lb/hp

Performance

Max speed at 16,000 ft	229 kt
77.5% power cruise at 10,000 ft	196 kt
65% power cruise at 10,000 ft	181 kt
Range at 77.5% power (10,000 ft)	1,388 nm
Range at 65% power (10,000 ft)	1,600 nm
Rate of climb at sea level Two engines	1,575 ft/min
One engine	230 ft/min
Take-off distance over 50 ft	2,367 ft
Landing distance over 50 ft	2,130 ft
Service ceiling Two engines	26,000 ft
One engine	10,100 ft

Engines

2 × Continental GTSI0-520-M developing 375 hp at 2,235 rpm, driving McCauley 90-in (229-cm) diameter, three-blade propellers.

TBO

1,200 hours as tested.
1,600 hours for modified engines.

Cessna 421 Golden Eagle

The ultimate in piston-engine light-twins

Background
Cessna, the biggest general aviation manufacturers in the business, entered the post-war light-twin market with their model 310. From it were inspired such aircraft as the delightful 340, one of the smallest of the pressurized twins, the 402 Businessliner (unpressurized but bigger), the similar but pressurized 414 Chancellor and, top of the range, the 421 Golden Eagle.

The first Golden Eagles were delivered during 1967 and although this high-performance twin is considerably more expensive than any of the aircraft described in this chapter (or for that matter any other piston-powered light twin) it has sold at an average rate of 150 units per annum. This is, perhaps, not surprising when one considers that the remarkable Golden Eagle can rival the performance of some turboprops.

Engineering and design features
The Cessna 421 is based upon a bonded wing with a glass-smooth wrinkle-free finish. Unlike the Titan described in the previous report the Golden Eagle does not have Fowler flaps. Instead, it uses good old-fashioned (but nevertheless effective) split flaps similar to those originally fitted to their model 310. Outboard of the engine nacelles the wet wing carries a total of 176 Imp/210 US gal (795 litres) of usable fuel, but as an option a tank may be fitted in the rear of one or both nacelles, increasing total fuel to 199 Imp/238 US gal (901 litres) and 223 Imp/268 US gal (1,014 litres) respectively.

When locker tanks are fitted these reduce the baggage capacity from 200 to 40 lb (91 to 18 kg) a side but you are still left with a 1,100-lb (499-kg) allowance within the fuselage – which is more than enough.

Similar 375-hp Continental engines to those in the Titan are used and these drive McCauley three-blade propellers through a reduction gear. The undercarriage and split flaps are actuated by hydraulic pumps, one on each engine, and wheel retraction time is normally 5 seconds. Even on one engine the wheels disappear in only 7½ seconds. The engines are spaced wide apart to allow the best possible clearance between propeller slipstream and fuselage sides.

Cessna 421 Golden Eagle, a piston-powered light twin with the performance of some turboprops.

To ensure a not-too-frightening V_{mca} a very large fin and rudder is provided. It does a good job; you can hold the aircraft straight, one engine inoperative, the other going full blast, at 80 kt. There is a fixed tailplane with separate elevators, the right-hand one having a large trim tab.

Baggage, 350 lb (159 kg), is carried in a nose bay; the nearby avionics bay (250 lb less the weight of optional radio equipment fitted); the two nacelle lockers already mentioned, and the rear cabin area which can accept another 500 lb. All this is of academic interest in an aircraft intended for, at the most, a pilot and six or seven passengers.

Six oval windows are let into each side of the cabin; the crew have somewhat bigger side windows and a large two-piece windscreen with storm windows that may be opened to give clear vision on the approach when the clouds are throwing it down in buckets.

Behind the wing on the left-hand side is the entrance door, top portion hinging up, lower section coming down to reveal a set of built-in steps. The Golden Eagle is a fine-looking aircraft when seen on the ground although, as a matter of personal taste, I have never particularly cared for the Cessna nose which suddenly comes to a point. The best-looking nose is on the smaller Cessna 340, but this is a minor detail. Overall the aircraft would grace any parking apron.

Cabin and flight deck
The 421 has an oval-section fuselage which is ideal for pressurization and a good shape for the occupants. It is almost 56 in (142 cm) wide – wider than a Beech King Air turboprop (see Chapter 6) – and the example flown had five luxurious armchairs plus a sixth which hides the toilet. Upholstery was in an opulent rust-coloured velour and the feeling of spaciousness completely disguised the fact that headroom at 51 in (130 cm) is little more than on the single-engine Beech Bonanza (see Chapter 4). Writing tables and other amenities make this a very civilized way to travel. The cabin is pressurized up to 5 psi and that is capable of providing the following conditions for passengers and crew:

Aircraft altitude (ft)	Cabin altitude (ft)
13,910	2,000
16,850	4,000
19,920	6,000
23,120	8,000
26,500	10,000
30,000	11,950

Although a 45,000 BTU combustion heating system is provided for very cold conditions, normally the cabin pressurization itself will maintain a comfortable interior even when it is well below freezing outside the aircraft.

The flight deck is a Cessna masterpiece. A long ledge panel, within easy reach of the captain's left hand, carries all circuit breakers other than those for the avionics and autopilot/flight director system which are on the right side-

wall. The ledge also carries the ignition, priming and starting switches, a voltmeter/ammeter and switches for the electric services.

The main instrument panel has few blank spaces. Starting from the left and moving right there is an annunciator with twenty-two warning lights. Below it are the pressurization controls and a contents gauge for the emergency oxygen bottles. The flight panel in this class of aircraft would include a flight director system using an ADI and an HSI in place of the usual artificial horizon and direction indicator. Air conditioning is handled on a small panel above the captain's right knee with the undercarriage switch next to it. The central area above the power quadrant is devoted to avionics and weather radar. To the right of the power quadrant is the 'follow-up' type flap control. The position indicator also shows maximum lowering speeds against the take-off and approach settings (176 and 146 kt respectively).

Cabin heat and windscreen defrost controls are at the bottom of the right-hand panel. Along the top of the main area, just below the glareshield, are seven engine instruments of the dual-finger type giving readings for the left and right engines. These are manifold pressure, rpm, a combined fuel-flow and fuel-consumed instrument sensibly calibrated in pounds weight (but not so sensibly named an 'Accru-Measure' by Cessna's desperate advertising boys and girls), EGT for accurate mixture adjustment, a three-finger instrument for each engine, reading oil pressure, oil temperature and cylinder head temperature and finally a two-finger fuel gauge indicating the amount of AVGAS in pounds within each wing. On the extreme right is a second flight panel for the copilot.

The power quadrant also carries trim wheels for all three axes and the autopilot control panel. Directly below, on the floor between the two pilots, is a pair of fuel selectors which may be set to OFF, LEFT MAIN or RIGHT MAIN. So crossfeed is affected by setting the fuel selector to the tank on the other side – how simple can you get?

The crew seats are the best I have seen. They adjust for height, leg reach and back rake. You settle in, press a button, then the seat adjusts pneumatically to give perfect lumber support. I find it hard to fault this flight deck in any way.

In the air
A LEFT/RIGHT priming switch, spring-loaded to the OFF position, is used during starting and the engines fire up readily. Visibility while taxiing is first class, particularly now that the wing-tip tanks of earlier models have been replaced by wet wings. Nosewheel steering is light and the very powerful brakes give a feeling of security. V_{mca} is 80 kt, V_r 95 kt and blue line, which is the same speed as best two-engine rate of climb, is marked on the ASI at 111 kt.

Like all turbocharged engines the throttles must be handled smoothly and progressively to avoid surging of the turbine unit. Despite this, the take-off in a Golden Eagle reminds you of a jet. It is remarkably quiet and acceleration is exhilarating; the aircraft does not hang around when you open the taps. We went up at a no-nonsense 2,000 ft/min and levelled out for the cruise checks at 12,000 ft. 73 per cent power gave a TAS of 210 kt for a fuel burn of 36 Imp/43 US gal (163 litres) per hour, but this is too low a level for a high-flyer like the Golden Eagle. For example, at 25,000 ft, 65 per cent power will fly you at 227 kt; at 75 per cent this remarkable piston-engine twin will beat some of the turboprops, returning 241 kt and producing 6.61 nm/Imp gal, 5.5 nm/US gal (1.45 nm/litre). For a high-performance twin, this is outstanding.

Flaps-up or down there is decisive elevator buffet as stalling speed is neared and the wings remain level after the 'g' break. Flaps-up I recorded a stall at 86 kt, flap 15

ABOVE: The comfortable passenger cabin is among the quietest of any piston-powered aircraft.

The wing is a rivet-free, bonded structure with a glass-like surface. Additional fuel tanks may be installed within the engine nacelles.

degrees removed 5 kt and flap 45 degrees reduced the speed to 75 kt. Control forces are just about ideal for a fast, light transport, with a firm rudder, less heavy elevators and light ailerons that handle faultlessly. All trimmers work well with the one for the ailerons a little oversensitive for my liking. This is one of the quietest piston twins I have flown.

The Golden Eagle's asymmetric performance is in a class of its own. With an engine feathered and the live one at 71 per cent power there was a 160 kt cruise at 12,000 ft. Fully loaded, the sea-level rate of climb on one engine is 350 ft/min, which is certainly above average.

The aircraft can integrate with turboprops and jets at busy airfields provided speed is reduced to 100 kt on short finals. Handling on the approach is beyond criticism. Power is brought back to the idle stop after the round-out and a short hold-off is made to ensure arrival on the mainwheels only. The nosewheel is lowered and then the powerful brakes may be used to curtail the landing roll. Considering its performance, the aircraft does not require much of a runway.

Capabilities

The Golden Eagle is offered in three versions: the basic bird with no frills, the series II which carries some equipment, and the Golden Eagle III with its full airways radio, de-icing and other extras which add 400 lb (181 kg) to the basic aircraft. Using this version as an example, equipped empty weight is 4,979 lb (2,258 kg) and if you subtract that from the 7,500 lb (3,402 kg) ramp weight there is a useful load of 2,521 lb (1,144 kg). Maximum zero fuel weight is 6,733 lb (3,054 kg), so 1,754 lb (796 kg) represents the limit that can be put in the cabin. Assuming the aircraft had the two locker tanks, maximum fuel would weigh 1,605 lb (728 kg), leaving 916 lb (416 kg) for, payload, say, a pilot and four passengers, plus 100 lb (45 kg) of baggage. You could fly a with-reserve 1,270 nm at 240 kt or 1,340 nm at 227 kt. The full cabin allowance of 1,754 lb (796 kg) would be enough for eight big adults and about 400 lb (181 kg) of baggage, with 767 lb (348 kg) available for fuel, ie, 107 Imp/128 US gal (485 litres); that is sufficient for more than 600 nm. So the Golden Eagle is not short of payload range. 250 lb (113 kg) of fuel must be burned off before a landing can be made following a maximum-weight departure. This amounts to about one hour in the air and in this class of aircraft I would not regard the landing weight restriction as much of a penalty to pay for the outstanding overall performance.

Verdict

The Cessna 421 Golden Eagle is faster than some turboprops with 500-hp engines. It can equal their rate of climb and range, but being piston-powered, is considerably cheaper to buy. Certainly it is very quiet and comfortable but like the Cessna Titan the 421 is powered by a pair of engines that require careful handling if premature removal

is to be avoided. Overhaul life has been increased from 1,200 to 1,600 hours and that would probably be enough for three to four years flying by the average business. Another problem is the unreliable supply of AVGAS in some parts of the world – but that cannot reflect on the excellence of the Golden Eagle, which is surely the ultimate light piston-engine twin.

Facts and figures

Dimensions

Wing span	41 ft, 1 in
Wing area	215 sq ft
Length	36 ft, 5 in
Height	11 ft, 5 in

Weights & loadings

Max ramp	7,500 lb (3,402 kg)
Max take-off	7,450 lb (3,379 kg)
Max landing	7,200 lb (3,266 kg)
Max zero fuel	6,733 lb (3,054 kg)
Equipped empty	4,979 lb (2,259 kg)
Useful load	2,521 lb (1,144 kg)
Seating capacity	8/10
Max baggage (with two locker tanks)	1,180 lb (535 kg)
Max fuel (with two locker tanks)	223 Imp/268 US gal (1,014 litres)
Wing loading	34.7 lb/sq ft
Power loading	9.9 lb/hp

Performance

Max speed at 20,000 ft	258 kt
75% power cruise at 25,000 ft	241 kt
65% power cruise at 25,000 ft	227 kt
Range at 75% power (25,000 ft)	1,270 nm
Range at 65% power (25,000 ft)	1,340 nm
Rate of climb at sea level Two engines (max operating)	1,940 ft/min
One engine	350 ft/min
Take-off distance over 50 ft	2,323 ft
Landing distance over 50 ft	2,293 ft
Service ceiling Two engines (max operating)	30,000 ft
One engine	14,900 ft

Engines

2 × Continental GTSIO-520-L developing 375 hp at 2,235 rpm, driving McCauley 90-in (229-cm) diameter, three-blade propellers.

TBO

1,200 hours as tested.
1,600 hours for modified engines.

Britten-Norman BN-2A Trislander

A piston-engine DC10

Background

Desmond Norman and his partner, the late John Britten, both of them graduates of the famous British de Havilland Technical School, set up shop in the crop-dusting business; soon they were designing and manufacturing their own spray equipment. One thing led to another and it was not long before the aviation world was shattered at the appearance of a small, ten-seat bush plane. It was simplicity itself, brilliant in concept and built like a brick outhouse. I am, of course, talking of the highly successful Britten-Norman Islander, more than 1,000 of which have been delivered world wide.

Desmond Norman, its codesigner, made the first flight on 13 June, 1965 and the Islander was certified two years later.

One of the reasons for the Islander's success is its ability to fly a pilot and nine customers out of the bush on the power of two Lycoming engines of either 260 or 300 hp. It was the right aircraft at the right time and at the right price. How about a bigger Islander? someone asked, believing that a need existed for a simple and inexpensive pilot-plus-seventeen-seat light transport. It was calculated that a suitably stretched Islander, using much of its STOL capabilities to lift a greater load, would require 780 hp. The obvious solution, you might imagine, would be to add a few yards to the Islander's fuselage, put in more seats,

Ingenious solution to an unusual problem – how to add a third engine to a low-slung aircraft (rear-mounted engine on the Britten-Norman Trislander).

install a brace of 390-hp engines and (probably the most difficult part of the exercise) think of a new name.

Nothing is ever straightforward in aviation and that particular plan spun in before it got started. First, if you move out of the 260-hp bracket into engines of higher power, simplicity goes down the plug and overhaul life takes a beating. Then, the higher-powered engines would entail a faster minimum control speed, require much stronger wings for the heavier motors, and raise doubts about the single-engine performance unless power was increased still further. How about four 200-hp engines? Well, the loss of an outer engine would certainly entail a high V_{mca}. So they looked at three engines on the basis that loss of one donkey during take-off would amount to a $33^{1}/_{3}$ per cent power reduction as opposed to a 50 per cent loss in the case of a twin. That looked attractive because the existing engines with their 2,000-hour TBO could be used in their original wing positions. But where to put the third engine? Obvious choice was in the nose but, like most high-wing designs, the Islander airframe sits low on the ground, too low for a nose-mounted propeller.

They chose the only other place, high up in the tail, stretched the fuselage in front of the wing to preserve the centre of gravity and called it the Trislander. In adopting an engine layout of this kind (two under the wings and one in the tail) the little firm of Britten-Norman set the fashion for mighty Douglas and Lockheed – their DC10 and Tristar L-1011, respectively. Although this chapter is entitled 'Piston-Engine Twins', the three-engine Trislander is most appropriately included here.

Engineering and design features

The wing is an adaptation of the twin-engine Islander's suitably beefed up to cater for an increase in weight from 6,600 lb to 10,000 lb (2,994 kg to 4,536 kg). The original Frise ailerons and electrically operated slotted flaps remain unchanged and the only external difference are the wing-tip tanks that increase the span by 4 ft and probably lift their own weight of fuel.

The two wing-mounted engines are slung close to the fuselage, an arrangement that ensures a low V_{mc} but at the same time contributes to cabin noise. A single, streamlined main undercarriage leg drops down from the rear of each engine nacelle to support double wheels; and the nose-wheel steers via the rudder pedals until brake is applied to tighten the radius of turn, at which point it automatically disengages to become free-castoring – an excellent arrange-

The flight deck is very convenient in most respects. The engine instruments are to the right of the captain's flight panel.

ment while parking.

A large, wide-chord fin supports the tailplane and its separate elevators at a position about two-thirds of the way up. At the intersection of the vertical and horizontal tail surfaces is the nacelle for the rear engine.

The very long fuselage is based on Islander frames, consequently it inherits the twin-engine plane's unique, narrow cabin with no central aisle between the nine rows of seats. How do you get in? It's like entering an old-fashioned charabanc with umpteen doors down the sides, in fact there are three on the right and two more at staggered positions on the left. There is a 28 cu ft (0.79 cu m) baggage compartment in the nose and a 25 cu ft (0.71 cu m) area behind the main cabin. This may all sound grotesque, yet from some angles the long slim Trislander, with its way-out, third engine banished up in the back room, does not look at all bad. I have seen a lot worse.

Cabin and flight deck

Entry and exit for crew and passengers alike is very easy. Being a high-wing design the fuselage floor sits close to the ground. The cabin is 43 in (109 cm) wide, 50 in (127 cm) high and 17 ft, 8 in (5.39 m) long from behind the pilot's seat to the rear luggage area, which adds another 8 ft (2.44 m) to the interior length. The nine double seats

span the full width of the cabin, making all five doors essential. Most of the passengers can enjoy an excellent view from their adjacent side windows and the seating is adequate for short journeys.

The flight deck is particularly well planned with related controls presented as they should be, in groups. For example, fuel and engine management run in a vertical area which starts with the fuel selectors, electric pumps and contents gauges along with the ignition switches in three pairs. That little lot is on a panel above the windscreen. Directly below and just under the glareshield are three mixture control knobs with, underneath them in the centre of the main instrument panel, three rpm indicators and three manifold pressure gauges, my only complaint being that they are not laid out in the same order as the engines: one engine behind the other two. A pedestal carries three throttles and three pitch levers followed by three carburettor heat controls – what could be better than that? Also on the pedestal is the flap control; press it down once and the flaps hit the take-off position. Press it again and they drop ready for landing. The flaps are similarly

In the air the Trislander, despite its unconventional layout, is not without appeal.

raised in two stages. There is a rudder trim wheel in the roof and an elevator trim on the pedestal.

Electric switches are grouped low down to the left of the pedestal and circuit breakers occupy the entire bottom right of the panel. Above them is the avionics installation and a standard flight panel is in front of the captain. The pilots' seats slide back and forth and the rudder pedals adjust for leg length so you can be a tailor's nightmare and still fly a Trislander in comfort. This is a businesslike, no-nonsense flight deck, not as glamorous as some but ideal in almost every respect. The view from up-front is superb.

In the air

With myself and eight trusting souls on board our weight was calculated to be about 8,500 lb (3,856 kg), so we could have carried another nine passengers if the need had arisen. Start-up is typical of any carburettor-type Lycoming and since the rear engine is a long way from the pilots, a red, the-back-one-has-failed warning light is included with the engine instruments. The brakes are first rate and nose-wheel steering is surprisingly light for an aircraft of this size. The flight manual quotes a V_2 (take-off safety speed) of only 66 kt but the manufacturers recommend that you rotate at 70 kt, go for the blue line 80 kt, raise the flaps and then climb at 100 kt unless you want the best climb gradient (for clearing trees, buildings and so forth), when 80 kt should be used.

The pilots sit well ahead of the engines, so noise level during the take-off and climb is not too bad but some of the 'ballast' sitting further down the cabin later described the experience as like being in a one-note disco. Initial climb rate was 1,100 ft/min (950 ft/min is claimed for a fully loaded Trislander).

At 4,000 ft, 75 per cent power (2,300 rpm/24.25 in manifold pressure) trued out at a by no means sluggish 152 kt, which is good for a simple, relatively low-powered, fixed-gear, seventeen-passenger light transport. Fully loaded the TAS would have been 149 kt but 67 per cent power reduces the noise level significantly at a cost in cruising speed of only 3 or 4 kt. Handling is somewhat firm, which is to be expected of this type of aircraft although I thought the ailerons were too heavy, possibly because the control yokes are little Mickey Mouse affairs that do not provide enough leverage.

Everything in the aircraft is easy to manage except for synchronizing the rear engine, and that takes practice. A synchroscope would be a great help. Lateral stability is above average but my pitch test was distorted somewhat because the 'ballast' had elected to sit towards the rear of the cabin and provide me with an aft centre of gravity. In the prevailing conditions two cycles were needed to regain a trimmed airspeed after a 15-kt displacement, something that would have been improved if more of the front seats had been occupied.

The stall is more docile than in many a light plane. Flaps-up the nose gently nodded, wings level, at 60 kt;

take-off flap (25 degrees) took 9 kt off; and with full-flap (56 degrees) the stall came at only 48 kt IAS. A few years previously I had flown a Trislander on the rear engine only with both wing-mounted propellers feathered. There were only three of us in the aircraft at the time but it was a hot day and we were flying at about 8,000 ft over Lanseria, South Africa. Renewing my acquaintance with the Trislander back home in England I thought the asymmetric handling was above average. They claim an engine-out climb of almost 300 ft/min at maximum weight and that is better than some of the light twins.

Recommended initial approach speed is 80 kt going for 75 kt over the hedge at high weights and only 65 kt when the aircraft is light. There is a nose-up trim as the flaps

Facts and figures

Dimensions

Wing span	53 ft
Wing area	337 sq ft
Length	48 ft, 2 in
Height	14 ft, 2 in

Weights & loadings

Max take-off	10,000 lb (4,536 kg)
Max landing	10,000 lb (4,536 kg)
Equipped empty (with pilot)	6,350 lb (2,880 kg)
Max zero fuel	9,700 lb (4,400 kg)
Useful load	3,650 lb (1,656 kg)
Seating capacity	18
Max baggage	700 lb (318 kg)
Max fuel	154 Imp/185 US gal (700 litres)
Wing loading	29.67 lb/sq ft
Power loading	12.82 lb/hp

Performance

75% power cruise at 8,000 ft	152 kt
67% power cruise at 9,000–10,000 ft	149 kt
Range at 75% power	540 nm
Range at 65% power	600 nm
Rate of climb at sea level Three engines	960 ft/min
Two engines	320 ft/min
Take-off distance over 50 ft	2,450 ft
Landing distance over 50 ft	1,950 ft
Service ceiling Three engines	13,600 ft
Two engines	3,000 ft

Engines

3 × Lycoming 0-540-E4C5 developing 260 hp at 2,700 rpm, driving Hartzell 80-in (203-cm) diameter, two-blade propellers.

TBO

2,000 hours.

are lowered but nothing that can't be handled while retrimming. During the landing it is essential to leave on approach power until after the round-out, otherwise the aircraft will sink heavily to the ground, but this is typical of larger aircraft.

Capabilities

Maximum take-off and landing weights are 10,000 lb (4,536 kg), and an airways-equipped example would have an operating weight (that is, equipped empty plus the crew – in this case one pilot) of 6,350 lb (2,880 kg), leaving a useful load of 3,650 lb (1,656 kg). There is a zero fuel limit of 9,700 lb (4,400 kg) so at least 300 lb (136 kg) of fuel must be carried. The remaining 3,350 lb (1,520 kg) can go in the cabin for high-load/short-distance operations. If you carry 17 passengers of average weight 170 lb (77 kg) each – and, of course, a pilot to fly the hardware – 760 lb (345 kg) of fuel may be uplifted (105 Imp/126 US gal or 477 litres). By settling for the 67 per cent power cruise and flying at 9,000 ft the Trislander would carry its 17 passengers over a with-reserve 380 nm at 149 kt. So loaded the Trislander will return 61.5 passenger nm/Imp gal, 51.26 nm/US gal or 13.54 nm/litre, which has got to represent value for money. Full tanks, taking 154 Imp/185 US gal (700 litres), could transport 15 passengers over a

range of 600 nm, but the aircraft has really been designed for short hops with either 17 passengers and their luggage, or up to 3,350 lb (1,520 kg) of freight when the seats have been removed (there is no landing weight restriction). Such a load could be flown for a credible 300 nm with enough fuel to provide a 100-nm diversion. When you consider it can fly these kinds of loads out of 2,500-ft airstrips on the power of three relatively small engines the Trislander is good news for operators who cannot afford turboprops.

Verdict

You will have noticed that I have not compared this aircraft with any other. That is because there are no other seventeen-passenger, three-engine piston designs on the market. Like its older brother, the Islander, the Trislander fills a unique position in general aviation. Economics are good, engineering is robust if a little basic and, for the short journeys intended, the Trislander is certainly fast enough. Noise levels are the main weakness of the design although this has not proved to be a problem in the field because passengers spend so little time in the aircraft. The Trislander is a brilliant aircraft in many respects, one that has proved itself operating under widely differing conditions.

HOW DO THEY RATE?

The following assessments compare aircraft of similar class. For example, four stars (Above average) means when rated against other designs of the same type.

★★★★★ Exceptional
★★★★ Above average
★★★ Average
★★ Below average
★ Unacceptable

Aircraft type	Appearance	Engineering	Comfort	Noise level	Visibility	Handling	Payload/Range	Field Performance	Cruising Speed	Economy	Single-engine Climb	Value for Money
Piper PA-44-180T Turbo Seminole	★★★	★★★	★★★	★★★★	★★★★	★★★★	★★★	★★★★	★★★	★★★	★	★★★
Beechcraft Duchess	★★★★	★★★	★★★★	★★★★	★★★★	★★★★	★★★	★★★	★★★	★★★	★★	★★★
Partenavia P68C	★★★★	★★★	★★★	★★	★★★	★★★★	★★★★	★★★★	★★★	★★★★	★★★★	★★★★
Piper PA-34-220T Seneca III	★★★	★★★	★★★★	★★★	★★★	★★★	★★★	★★★	★★★	★★★	★★	★★★★
Cessna T303 Crusader	★★★★	★★★	★★★★	★★★★	★★★★	★★★★	★★★★	★★★★	★★★	★★★★	★★	★★★
Piper Aerostar 602P	★★★★★	★★★★	★★★	★★★	★★★★	★★★★	★★★★	★★★	★★★★★	★★★★★	★★★★	★★★
Beechcraft Baron 58P	★★★★	★★★★★	★★★	★★★★	★★★	★★★★	★★★	★★★	★★★★	★★★★	★★	★★★
Piper PA-31 Navajo	★★★	★★★	★★★★	★★★	★★★★	★★★★	★★★★	★★★★	★★★	★★★★	★★★	★★★★
Cessna 404 Titan	★★★★	★★★	★★★★	★★★★	★★★★	★★★★	★★★★★	★★★★	★★★	★★★★	★★	★★★
Cessna 421 Golden Eagle	★★★★	★★★	★★★★	★★★★★	★★★★	★★★★	★★★★	★★★★	★★★★	★★★★	★★★★	★★★
Britten-Norman BN-2A Trislander	★★★	★★★★	★★★	★★	★★★★	★★★	★★★★★	★★★★	★★★★	★★★★★	*★★★	★★★★

*Two-engine climb

6. TURBOPROP AIRCRAFT

After many decades of experience and development, the piston engine has attained a remarkable degree of smooth and reliable performance in automobiles but, in the main, car engines are not suitable for aircraft use. Propeller-tip speeds may not be allowed to approach that of sound, otherwise shock waves are formed, power is dissipated in making a lot of noise and that in turn can be relied upon to annoy the neighbours.

How do you ensure low propeller rpm? Well, the simple answer is build slow-running engines. But then we hit a problem because, power for power, slow engines are bigger than high-revving ones and on average even a light-plane motor is twice the volumetric capacity of an automobile engine. For example, the 112-hp Lycoming engine, adopted almost exclusively by two-seat trainer manufacturers, has the swept volume of car engines developing more than 200 hp – but at twice the speed.

Such engines, relying upon size to develop power at low rpm, offer good reliability, and so they should, because simplicity is the keynote. The propeller is bolted directly to the end of the crankshaft, there are no complications and you can expect a TBO of 2,000 hours. All this is fine for outputs of up to 300 hp or so but when more horses are required, the direct-drive engine becomes big and heavy – a lump of metal that can be difficult to install.

The alternative is to run the engine faster and keep the propeller speed within bounds by driving it through reduction gears. Reduction gears do not sound much of an engineering problem, but the trouble with piston engines driving a propeller through a lot of teeth is that vibrations can be set up and magnified. Put a child on a swing and continue applying the same gentle push. Soon the screaming youngster will be almost two parts around a loop. So it is with reduction gears. Harmonic vibrations can rapidly build up to the point where serious damage occurs. The answer lies in arranging some form of spring drive to cushion the interaction of piston-engine/propeller vibration on the teeth of the gears; but few of the lower-powered aero engines with reduction gears have been as reliable as their direct-drive relations.

There is another problem with piston engines, relating to corporate aircraft – namely, the poor availability of AVGAS. The petroleum companies have created this situation by their grand claims of 'special additives' for aviation fuel. There was a time when all aero engines ran on good-quality motor spirit, then around 1946 the 'additive' craze took a hold, the aviation fraternity was brainwashed into believing there was something special about light plane fuel (partly to justify an increase in price) and the habit was formed. Now that air transport fleets are 99.9 per cent turbine powered, the demand for AVGAS has shrunk, the petroleum companies can no longer be bothered to produce small quantities for general aviation and they dare not say 'buy it from the garage' because that would amount to admitting we have had our legs pulled over the years. To some extent the situation is being taken out of their hands because, among other establishments, CSE Oxford, the biggest professional pilot school in Britain (also one of the largest in the world), is operating its single-engine fleet on motor fuel – MOGAS to its friends. And the UK Civil Aviation Authority is the world leader in supporting the trials that have led to this interesting experiment.

Having said all that, many countries insist on filling general aviation aircraft with AVGAS and herein lies the problem, because often the most insistent ones are in no position to supply it. There have been cases of piston-engine twins being stuck in parts of Europe until some kind soul flew in supplies. All over the world you will find places where AVGAS is not obtainable. On the other hand, jet fuel (AVTUR) is the lifeblood of the airlines and you can buy that at any airfield where the local banana run represents the event of the day.

So the advantages of turboprop engines are availability of fuel, hardly any vibration (things go round and round, not up-stop, down-stop) and better reliability than even the best piston engines. The disadvantages are very high initial and overhaul cost. A small turboprop engine delivering 300–400 shaft horsepower (shp) might cost four times the price of a 375-hp piston engine. To offset this, TBO of a turboprop engine varies between 3,000 and 6,000 hours according to type, whereas the geared piston motors run for only 1,400–1,600 hours. Fuel consumption is very much higher for turboprop engines, particularly while flying at lower levels, but to some extent this is redressed by the fact that in most parts of the world AVTUR is only two-thirds the price of AVGAS.

Many well-informed folk in aviation believe that turboprops will eventually replace piston engines in all but the microlight and powered-glider categories. Before that happens, and personally I would welcome the introduction of gas turbines on a large scale, the price has got to be

Application of reverse thrust. British Aerospace Jetstream 31

brought within reasonable bounds. Fortunately, one or two new manufacturers are working hard at the problem and their efforts may inspire a new range of light turboprop engines costing half the price of existing units.

What do we require of a turboprop? All the talents expected of the piston-engine twins listed at the beginning of the previous chapter plus:

- Proper ice protection cleared for flights into known icing conditions.
- Low noise levels. A few of the turboprops have been a let-down in this respect and there is no excuse for that in an expensive aircraft.
- An adequate cabin. Smallest turboprop engines develop 420 shp (although they may be flat-rated to 300–320 shp when all the power is not required) and they range in power up to 900 shp or more. Being small and light it is tempting to fit high-powered turboprops into relatively small airframes. The customer is nevertheless entitled to a good cabin as part of the deal.

- The best possible fuel economy. Remember, turbines are thirsty at low levels.
- They must offer significant performance improvements over piston-engine aircraft to justify their relatively high cost.
- Since the initial cost of turboprop engines and their subsequent major overhaul is high, only those engines offering a good TBO should be installed by the airframe manufacturers. The current 3,000–3,500 hours must be improved upon by the engine companies.

In the eight airtests that follow, turboprop terminology will be used and readers who are not conversant with this aspect of aviation are advised to refer back to the 'Engine terminology' section in Chapter 1, which explains briefly the various controls and instruments that will be unfamiliar to piston-engine pilots (see pages 12–13).

Pilatus PC-6 Turbo Porter

The Super-STOL from Switzerland

Background

Pilatus Aircraft of Stans, Switzerland, is part of the Oerlikon-Buhrle group, which is famous for anti-aircraft guns, so at first sight there might appear to be a conflict of interest here. Pilatus are not new to building aircraft. At one time they designed a trainer which was to prepare German and other pilots destined to fly Messerschmitt 109 fighters. In more recent times they have produced the very successful PC-7 turboprop military trainer/light ground-attack aircraft. The subject of this airtest is a very interesting utility aircraft that has sold in considerable numbers on a worldwide basis: the PC-6 Turbo Porter. The original Porter was piston powered, but since the turboprop version first appeared in 1959 some 920 have been delivered (up to the beginning of 1984).

Pilatus claim that the Turbo Porter can be used for passenger or cargo transport, aerial survey, search-and-rescue, supply dropping, air ambulance work, air photography, rain provoking by cloud seeding, parachute training, fire-fighting (water-bombing) target and glider towing, and crop spraying or dusting. This is some list and one is tempted to read their claims with a double-take. But the Turbo Porter is actually being used in all these roles – and a few more besides – on a day-to-day basis. It also has a special talent: the ability to operate in and out of strips that would defy the field performance of any aircraft other than a helicopter.

Engineering and design features

For a single-engine aircraft the Turbo Porter is a monster, with a span nudging 50 ft, a great long nose – yards of it (literally) – and wheels that from a distance make it look like a little boy with his dad's shoes on. The light weight of the PT6A-27 free-turbine engine, which is flat-rated to 550 shp, has demanded that it be installed well forward of the wing. There is no pandering to aesthetics or fashion on this aircraft. Everything is brutally functional with slab sides, square-cut flying surfaces and little attempt at streamlining.

Mechanically operated, double-slotted flaps of large area take up half the span of the strut-braced wing which contains the 142-Imp/170-US-gal (644-litre) fuel system. The square fin sits bolt upright and, like the elevators, its rudder has an overhanging horn balance reminiscent of those Junkers 52 transport planes used by the Luftwaffe during the Second World War. There are no pilot-adjustable trim tabs on the elevators. Instead, the incidence of the tailplane is altered from the flight deck by an actuating gear (a screw-jack) of the type that used to be popular in the 1920s and early 30s.

The PC-6 is one of the biggest taildraggers flying, and the tailwheel assembly is a substantial piece of engineering that is steered through the rudder pedals. There is a control on the flight deck for locking the tailwheel during the take-off and landing. The mainwheels, massive affairs almost 3 ft in diameter, are attached to 'V' struts, their upper ends hinging on the centre-line of the fuselage underside, the lower ends terminating in wheel axles. Springing is provided by long oleo struts which attach to near the wheel axles, their top ends being anchored high up on the fuselage sides.

The Jimmy Durante of the aviation world – the long nose of the Pilatus PC-6 Turbo Porter.

The fuselage has a shape that only its mother could love. It is a great, flat-sided edifice with abrupt contour changes. On the right-hand side of the cabin area is a very large sliding door measuring 41 in (104 cm) high and 62 in (158 cm) long. Directly opposite on the left-hand side is a pair of double doors that hinge apart to reveal an opening of the same dimensions. Both sliding and hinged doors are fitted with windows, the likes of which I have never seen on any other aircraft. In style I would describe them as modified Gothic. At the rear of the cabin area a large

circular porthole is let into each side and two small doors are provided for the pilots. Pilots must indulge in a little mountaineering up the sides of the aircraft to enter the flight deck, but these doors are essential because the crew seats firmly divide the flight deck from the cabin area. There is no way through from one part to the other.

Under the main cabin area is a trap door let into the

floor and when the PC-6 is used to drop supplies, up to 660 lb (299 kg) can be launched into space by the pilot, without having to leave his seat. He can even close the doors afterwards. The floor hatch is also used when vertical cameras are employed; and when a water tank is installed for fire fighting, 175 Imp/210 US gal (795 litres) may be carried – sufficient to extinguish an area measuring about 60 × 400 ft.

Engineering features could fairly be described as of good quality but crude design.

Cabin and flight deck
The Turbo Porter flown for this test is operated by one of the leading parachute schools in Britain, consequently there were no seats in the cabin area. So when the big door on the right was slid back, it was almost like opening the hangar. The cabin is about 46 in (117 cm) wide, 50 in (127 cm) high (with a ceiling that lowers by 4 in (10 cm) at the rear) and 8 ft, 4 in (2.54 m) long from behind the crew seats. When used for carrying passengers the standard layout is two crew seats (pilot and passenger or, if required, two pilots) and six in the cabin. There is adequate room for such a load. Alternatively, three rows of three seats can be installed except in Britain, where the airworthiness authorities limit the cabin to eight passengers in addition to the crew seats.

The flight deck is as unconventional as the rest of the aircraft. A large handle in the roof is wound to lower the flaps, their position indicator taking the form of a marked probe which extends from the left-hand wing leading edge as they come down. Take-off position is 28 degrees and the maximum setting (used for landing) is 38 degrees. Concentric with this handle is a smaller one for the tailplane adjustment, which replaces the usual elevator trim. I found it heavy and awkward to use in flight and prospective purchasers of the Turbo Porter are strongly recommended to pay a little more and have the optional electric trim. Low down on the left of the captain's seat is a rudder trim, and nearby is the control for locking and unlocking the tailwheel. The seats are reasonably comfortable, although firmly fixed to the floor, but the rudder pedals adjust for leg reach.

There is a wide, shallow instrument panel with the usual flight instruments on the left, the engine instruments and starting controls in the centre, and radio and electrics mixed up with fuel contents and oil temperature/pressure gauges on the left. Across the full width of the flight deck is a shelf that runs along the bottom of the instrument panel – a useful feature for placing maps, airways manuals and so on. In its centre is a small power quadrant with the levers arranged in nonstandard order. From left to right they are the propeller control, the power lever and the fuel condition lever.

Pilatus should devote some time to bringing this flight deck out of the 1940s. Unrelated controls and readouts need regrouping, the trim control is the worst I can remember and the tailwheel lock needs intimate knowledge before it will engage. We used to put up with this kind of eccentricity in the days when helmets and goggles were part of the kit. Aviation has come a long way since then and the Turbo Porter with its remarkable performance, which is unequalled by any other aircraft in its class, deserves a face-lift.

In the air
Although the nose slopes down there are 10 ft of it in front of the windscreen so, being a taildragger, it is essential to swing left and right to clear the view ahead while taxiing. The PC-6 fired up without hesitation (turbines start more easily than piston engines) and we moved off. The aircraft steers with the rudder pedals through 25 degrees left and right of centre but for tighter-radius turns a little brake in the required direction allows the tailwheel assembly to disconnect and free-castor. Steering is good and so are the brakes.

For the first take-off I would only have the demonstration pilot on my right, there being no seats in the back in which to strap human ballast during stalling. I was advised to line up on the runway, lock the tailwheel, open up to maximum torque on the brakes then let her fly off without raising the tail. Looking back on the experience I still find it hard to believe. Within seconds of brakes-off we were climbing at 2,000 ft/min and I estimate we were at the 50-ft point over the airfield in a distance of no more than 500 ft despite there being very little wind.

Flaps-up and power off I tried a stall. There was no 'g' break, just a nose-high sink at 45 kt indicated. Full-flap produced a gentle nod of the nose at 40 kt. I then returned to pick up some passengers for a load test. Eight parachutists arranged themselves on the floor and I naturally expected there to be a much longer take-off roll. This time I opened up the power lever and let the aircraft roll. We had left the ground in 5 seconds and were climbing at 1,300 ft/min with a full load.

For cruise control you set the propeller at 92 per cent and adjust the power lever to attain 134 kt IAS. The torque meter, by the way, is calibrated in pounds pressure. It is time the manufacturers agreed on a standard system; personally I favour percentages of maximum torque as used on the BAe Jetstream (see airtest later in this chapter). At 2,500 ft the fuel burn at 137 kt TAS was 268 lb/hour. Naturally the aircraft becomes more economical when flown at 10,000–12,000 ft. Although there were no furnishings in the cabin, noise level was remarkably low.

Being like a big Cessna 152 (see Chapter 2), the pilots sit more or less in line with the wing leading edges and it is necessary to lean well forward to avoid blindspots while

OPPOSITE ABOVE: The flight deck is in need of re-vamping and some of the controls could be easier to use.

OPPOSITE BELOW: As a utility aircraft the Turbo Porter can accomplish the near impossible in terms of weight lifting and its ability to operate from tiny airstrips.

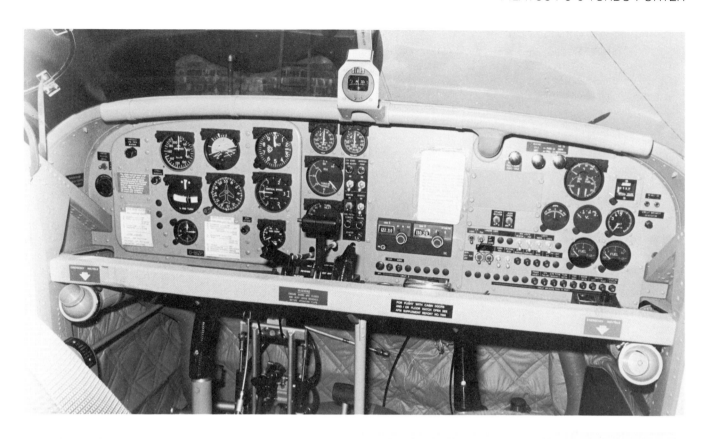

turning. Otherwise visibility in the air is acceptable. It is essential to use rudder while turning if balance is to be maintained, and power adjustments demand constant attention to the rudder trim. The elevators are firm but effective and so is the rudder, but the ailerons are the best feature from the handling point of view. The trimmer is certainly difficult to live with. Stability in pitch was not easy to check because the boys and girls were fidgeting while trying to sit comfortably on the floor. Lateral stability is neutral with little or no righting tendency.

I was talked into an obstacle-clearance STOL landing. The idea is to approach with full-flap, settle the speed at 60 kt and hold 1,500 ft until the airfield boundary appears under the nose and it looks as though you will be landing in the next county. The power lever is brought back to the idle stops and as the three-blade propeller turns itself into a drag-producing disc the nose must be lowered to the dive-bombing attitude if 60 kt is to be maintained. I actually had to add a little power to prevent undershooting!

Near the ground, there was a big change of attitude as I made the round-out, the 60 kt drained away almost instantly and before I could catch up with the aircraft we were landed on all three wheels. I applied reverse thrust, hit the brakes, and stopped in a ground roll of not more than 50 yards. All very impressive.

Capabilities

For public transport and most other duties the Turbo Porter is certified at a maximum weight of 4,850 lb (2,200 kg). An average equipped empty weight would be 2,770 lb (1,257 kg) and that leaves a useful load of 2,080 lb (944 kg). Full fuel (1,122 lb/509 kg usable) allows a payload of 958 lb (435 kg), which is enough for six occupants including the pilot of average weight 159 lb (72 kg). You could fly such a cabin for 470 nm allowing for a 95-nm diversion. With ten 160-pounders on board the Turbo Porter will fly for about 200 nm.

For crop spraying the PC-6 may be operated at up to 6,100 lb (2,767 kg), when the useful load becomes a remarkable 3,330 lb (1,511 kg) – considerably more than its empty weight. Landing weight remains 4,850 lb (2,200 kg), but that is no hardship, because the load will have been dumped before returning. The landing gear has been drop-tested to 9 ft/sec, so you would have to try hard if you wanted to break the aircraft. Naturally both rate of climb and cruising speed are reduced when flown at the overload weight but this is of little consequence, having regard to the nature of agplane operations.

Verdict

The Turbo Porter may be seen on skis performing the impossible off ice and snow, on floats and on wheels. They police the desert in the Middle East, fly big loads to impossible strips all over the world and generally do an excellent job. I can forgive its ugly appearance but need the engineering be so crude? And why not bring the flight deck up to present-day standards? The Swiss have a reputation for being thorough but I feel bound to accuse the Pilatus team of resting on its laurels with the Turbo Porter. It has unmatched capabilities and fully justifies the manufacturer's claim that it is a 'Super-STOL'. But it is high time the design was refined.

Facts and figures

Dimensions

Wing span	49 ft, 10 in
Wing area	310 sq ft
Length	36 ft, 1 in
Height	10 ft, 6 in

Weights & loadings

Max take-off Normal	4,850 lb (2,200 kg)
Overload	6,100 lb (2,767 kg)
Max landing	4,850 lb (2,200 kg)
Equipped empty	2,770 lb (1,257 kg)
Useful load Normal	2,080 lb (944 kg)
Overload	3,330 lb (1,511 kg)
Seating capacity	10/11
Max fuel	142 Imp/170 US gal (644 litres)
Wing loading Normal	15.64 lb/sq ft
Overload	19.67 lb/sq ft
Power loading Normal	8.81 lb/hp
Overload	11.09 lb/hp

Performance

Economic cruise Normal	130 kt
Overload	115 kt
Range at 130 kt and 4,850 lb (2,200 kg)	470 nm
Rate of climb Normal	1,270 ft/min
Overload	830 ft/min
Take-off distance over 50 ft	770 ft
Landing distance over 50 ft	715 ft
Service ceiling Normal	28,000 ft
Overload	20,500 ft

Engine
Pratt & Whitney of Canada Ltd PT6A-27 free-turbine flat-rated to 550 shp, driving a Hartzell 101-in (2.57-m) diameter, three-blade, reverse-pitch propeller.

TBO
3,500 hours plus 500 hours extension 'on condition'.

Cessna 425 Corsair/Conquest I

First step up to turboprop flying

Background

No doubt provoked by the uncertain nature of AVGAS supplies in some parts of the world and the fact that Piper had for some time been turning out turboprop versions of their Navajo piston twin (see Cheyenne III test later in this chapter), Cessna performed a similar exercise on their Golden Eagle (see Chapter 5). It was easier for them than for Piper, because in the Golden Eagle Cessna had a pressurized aircraft. That is always a good starting point when you have ambitions to build a plane capable of cruising at 16, 18, 20 or more thousand feet without subjecting the inmates to the indignity of having to wear oxygen masks (at least it stops them smoking or biting their nails).

There are not that many private owners flying aircraft in the Golden Eagle class. Price is one reason but complexity of systems and the very nature of such an advanced aircraft really demand the services of a professional pilot. However, Cessna have been at pains to make their turboprop Golden Eagle, the 425 Corsair, as simple as possible. They claim that it can be flown by the accomplished amateur and no doubt that is true – provided the amateur adopts a professional attitude towards handling, flight deck procedure, checks vital and not so vital, instrument flying, navigation and engine management. Like it or

not, by the time you are ready to strap yourself into a twin-turboprop, pressurized, scaled-down airliner, professionalism is even more important than finding the money to pay the bills.

To add a little confusion, in 1982 the Cessna marketing people changed the name of the Corsair to Conquest I, although it bears no resemblance whatsoever to their bigger turboprop, the Conquest II. They have also strengthened the undercarriage and tyres so that the maximum weight can be increased by 400 lb (181 kg).

Engineering and design features

Much of the airframe is Golden Eagle. The wing has 18-in (46-cm) extensions outboard of the ailerons, otherwise, apart from an increase in fuel capacity from the Golden Eagle's 223 Imp/268 US gal (1,014 litres) to 305 Imp/366 US gal (1,385 litres), there are few changes. Turboprop engines are small, compact and light for their power, so to preserve the centre of gravity the PT6A-112 free-turbine power units, which are flat-rated to 450 hp, are carried

Cessna 425, originally known as the Corsair but now re-named the Conquest I. Its nearest competitor is the Piper Cheyenne I.

well forward of the wing leading edge on long, slim nacelles that terminate in streamlined cones behind the wings. There are no nacelle lockers as in the Golden Eagle so maximum baggage is only a mere 1,100 lb (499 kg), which is enough for 25 passengers at the standard airline rate. The engines are slightly closer set than the Golden Eagle's and this no doubt compensates for the more powerful yawing moment that would result from that extra 75 hp a side during asymmetric flight. In any case, the Corsair has an absolutely massive fin and rudder. Any bigger and it would spoil the lines of this attractive little turboprop.

The tailplane is of greater span than that of the Golden Eagle. It has pronounced dihedral, lifting it out of the main areas of slipstream. It is also claimed to reduce Dutch roll tendencies. From the tailcone forward to the windscreen the fuselage is virtually that of its piston-engine mother, but the Corsair has a slightly shorter nose of more attractive profile. Up to 350 lb (159 kg) baggage can be carried in the nose bay; up to 250 lb (113 kg) in the avionics bay, subject to the amount of radio equipment installed; and up to 500 lb (227 kg) at the rear of the cabin. Window arrangements are the same as those of the Golden Eagle and so is the two-piece door with its built-in airstairs.

Each engine has a hydraulic pump for the undercarriage system, with electrically controlled selectors. The split flaps are actuated by an electric motor, and a follow-up linkage ensures they go where the flap lever tells them to.

Electric power is supplied by 28-volt DC generators, which in common with standard turboprop practice, double as starters for the engines. There is a 24-volt, 39-amp ni-cad battery complete with a 'battery overheat' warning system. AC current for the avionics is supplied by a dynaverter.

Pressurization is similar to that of the Golden Eagle, with a maximum differential of 5 psi giving a sea-level cabin up to aircraft altitudes of over 10,000 ft (see table of aircraft/cabin altitudes on page 155). Cabin heat is provided by a controllable system that utilizes bleed air from the engine compressors. An additional electric fan-heater is also installed.

The engines have their own ice protection but de-icing boots may be fitted to the wing, tailplane and fin leading edges. This is essential in an aircraft capable of climbing through the weather.

While the Corsair is not built to big-aircraft standards, by nature a pressurized turboprop must be made to higher levels of engineering than, say, a light tourer. The 425 is nicely designed and beautifully finished. The bonded wing has a surface like an ice rink.

Cabin and flight deck

You enter the bomber in true airliner style, up the stairs and into the 56-in (142-cm) wide cabin. The Corsair tested had four armchairs in the passenger area, a corner seat and another one in the baggage section making eight seats including the two up front for the crew. However, the aircraft is at its most comfortable when flown as a six-seater, when it will provide very civilized transport in a relatively spacious cabin. From the instrument panel to the rear baggage area is the best part of 16 ft. The seats in the example flown are covered in a very attractive blue tweed and it would be no hardship to contemplate a long journey in such beautiful surroundings.

Flight deck layout is very similar to the Golden Eagle's except for the engine instruments, which are replaced by those tiny dials that come with turboprops. They are mounted in two vertical columns (left engine and right engine) to the right of the captain's flight panel. In pairs, top to bottom, they read: engine torque, propeller rpm, interstage turbine temperature (ITT), gas generator rpm in percentages of maximum (just as well — the compressor does 37,500 rpm at full chat), fuel flow in pounds per hour, and combined oil-pressure/oil-temperature gauges. An annunciator panel is high up on the centre of the panel and the electrics are controlled on a ledge by the captain's left hand that's rather larger than the one on the Golden Eagle.

The central pedestal carries the engine power levers, propeller controls and fuel condition levers with two positions only: CUT-OFF and RUN. Nearby are the three trimmers (elevator, aileron and rudder) and the autopilot control box. There is also a small panel for the propeller synchronizer. Being such a wide fuselage there is ample room for two flight panels with three stacks of radio between. The flight deck is attractive and well planned.

In the air

With 1,400 lb (635 kg) of fuel in the wings, some baggage, four trusting souls in the passenger area, Cessna's demonstration pilot in the right-hand seat and yours truly on the captain's throne the aircraft stood on the ground weighing 7,600 lb (3,447 kg), and that left 600 lb (272 kg) in hand. So we could have taken 2,000 lb (907 kg) of fuel with such a cabin load, had the journey required it. Free-turbine engines are started and shut down with their propellers feathered, cutting out unwanted slipstream and stopping propeller rotation more quickly so that people can disembark in safety at the end of the trip. Otherwise, since the propeller is not mechanically linked to the compressor unit, it might waft around for quite a long time, unlike propellers on piston engines which cease rotation almost immediately following shut down. So with the ignition and fuel booster ON, I engaged the starter switch, and when 12 per cent compressor rpm showed, the condition lever was moved to the RUN position, allowing fuel to enter the spray nozzles. Almost immediately a rise in ITT confirmed light-up and the engine rpm began to accelerate. At 52 per cent the engine is self-sustaining, the start switch is turned off, the power lever is opened to 68 per cent and then the generator switch is turned on, allowing the starter motor to begin putting energy back into the battery. Other than minor variations in the percentages mentioned, this starting procedure applies to the other turboprops in this

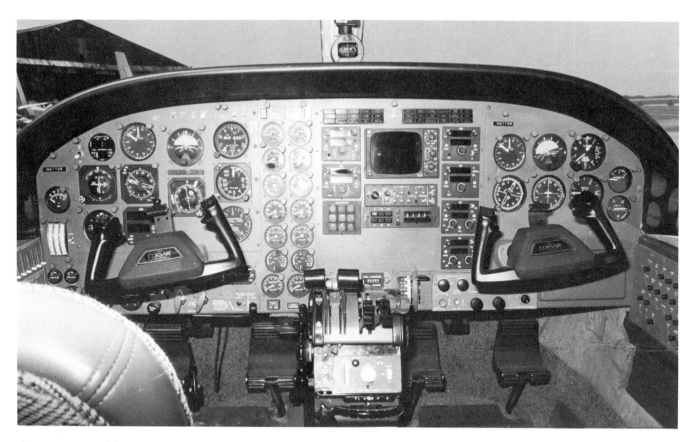

chapter powered by PT6 engines.

At this stage the feathered propellers are drifting slowly around and will continue to do so until the propeller levers are advanced to the normal-range position.

Nosewheel steering is remarkably light and the brakes are first class. The normal tendency for a turboprop to taxi like an express train is discouraged by lifting the power levers and bringing them back into the Beta mode where little or no thrust is provided by the near-neutral blade angle. The recommended technique for take-off is to hold on the brakes, move the power levers forward, let the engines stabilize at 600 ft: lb torque, then release the anchors and advance the power levers at 1,200 ft: lb as the aircraft starts to gallop. The torque meters are red-lined at 1,477 ft:lb. Minimum control speed is 90 kt, V_r 98 kt and blue-line speed 111 kt. All these speeds are passed and left behind as the Corsair accelerates straight and true down the runway, lifting off within seconds. A comfortable cruise/climb is 150 kt and at that speed the VSI showed a lusty 2,000 ft/min ascent. The engines can maintain their flat-rated 450 hp to almost 18,000 ft, so rate of climb remains constant to that level. The folk sitting in the gin-and-tonic seats expressed their satisfaction at the low noise level during our departure.

At 18,000 ft the recommended maximum cruise power (1,900 rpm/1,237 ft:lb torque) gave us a cracking 267-kt TAS – about 6 kt better than the figure quoted for a fully loaded example. Fuel burn was 260 lb (118 kg) per engine.

Cessna flight decks are among the best in general aviation. Engine readouts are in two columns just left of centre.

Maximum range cruise entails selecting 1,900 rpm and altering the power levers to adjust torque according to aircraft weight, altitude and outside air temperature. For example, at 18,000 ft you should go for 207 kt TAS under ISA conditions at maximum weight – fuel burn will be 171 lb (78 kg) per engine, which is not excessive for a turboprop. Noise levels in all parts of the aircraft are very low.

General handling is above average with good stability to match. Visibility is better than from any of the pressurized twins. However, the stall (90 kt clean, 84 kt wheels and flaps down) provoked the mother and father of a wing drop and while I acknowledge that there are no recovery problems, Cessna should improve this. Asymmetrics present no difficulties. Single-engine service ceiling is, at 20,000 ft (while at average weight), equal to some twins with both engines running, and the maximum-weight, single-engine climb of 450 ft/min, with at least 600 ft/min at average operating weights, is a great confidence builder, although turboprop engines are rarely known to stop during the take-off.

On the approach the aircraft is rock steady with excellent speed stability. Threshold speeds vary from 102 down to 95 kt according to weight, and the elevators remain active to the last. You round-out, bring the power levers fully back to the stops, let the mainwheels touch, lower the nosewheel, then lift the power levers and move them back into reverse thrust to shorten the landing roll by about 20 per cent.

Capabilities

The well-equipped example tested has an equipped empty weight of 5,100 lb (2,313 kg), leaving a useful load of 3,175 lb (1,440 kg). Maximum fuel weighs in at 2,452 lb (1,112 kg), leaving 723 lb (328 kg) for pilot and, say, 3 passengers plus baggage, when it will fly a with-reserve 1,265 nm at 26,500 ft, cruising at 257 kt, or over 1,600 nm at 210 kt (30,000 ft). A pilot and 7 passengers plus 400 lb (181 kg) of baggage, totalling 1,760 lb (798 kg), would leave 1,415 lb (642 kg) for fuel, which would fly you 625 nm at over 250 kt or 900 nm if 210 kt is fast enough. Field performance is similar or slightly better than the Golden Eagle. The 200-lb (91-kg) difference between maximum and landing weights is perfectly acceptable in an aircraft of this type, representing less than thirty minutes fuel burn at average speeds.

The Conquest I version, with its 400-lb (181-kg) increase in maximum weight, now offers another 374 lb (170 kg) of useful load and that is enough for two more passengers when maximum fuel is carried, making it a full tanks/six-seat aircraft.

Verdict

The Corsair is very much dearer to buy than a Golden Eagle, but to offset that, its engines have more than twice the overhaul life of the geared Continentals. Here is a brief comparison of the two aircraft. The current Conquest I development of the Corsair compares even more favourably.

	421 Golden Eagle	425 Corsair
Max ramp weight	7,500 lb (3,402 kg)	8,275 lb (3,754 kg)
Equipped useful load	2,521 lb (1,144 kg)	3,175 lb (1,440 kg)
Single-engine climb	350 ft/min	450 ft/min
Max cruise	241 kt	257 kt
Range at max cruise	1,270 nm	1,265 nm

On the face of it the Corsair offers little in terms of additional performance over its piston-powered relation but there is no comparison in engine life and reliability. Also you can get turbine fuel anywhere.

The 425's fine proportions show up well in the air.

The Corsair offers a bigger cabin than the Piper's little turboprop, the Cheyenne I, it is a little faster and it has more range.

As a first step into the world of turboprops the Corsair is a beautiful-to-look-at, splendid-to-ride-in, easy-to-fly private airliner. And the icing on the cake? Cessna are offering retrofit kits to bring existing Corsairs up to Conquest I standard. Modifications include parts for the undercarriage and ten-ply tyres in place of the original eight-ply equipment.

Facts and figures
(Corsair version)

Dimensions

Wing span	44 ft, 1 in
Wing area	225 sq ft
Length	35 ft, 10 in
Height	12 ft, 7 in

Weights & loadings

Max ramp	8,275 lb (3,754 kg)
Max take-off	8,200 lb (3,720 kg)
Max landing	8,000 lb (3,629 kg)
Max zero fuel	6,740 lb (3,057 kg)
Equipped empty (as flown)	5,100 lb (2,313 kg)
Useful load	3,175 lb (1,440 kg)
Seating capacity (with crew)	8
Max baggage	1,100 lb (499 kg)
Max fuel	305 Imp/366 US gal (1,385 litres)
Wing loading	36.45 lb/sq ft
Power loading	9.1 lb/hp

Performance

Max speed (max continuous power) at 17,700 ft	264 kt
Max cruise at 26,500 ft	257 kt
Max range cruise at 30,000 ft	210 kt
Range at max cruise power	1,265 nm
Range at max range power	1,646 nm
Rate of climb Two engines	2,027 ft/min
One engine	450 ft/min
Take-off distance over 50 ft	2,341 ft
Landing distance over 50 ft	2,145 ft
Service ceiling Two engines	34,700 ft
One engine	18,500 ft

Engines
2 × Pratt & Whitney of Canada PT6A-112 free turbine developing a flat-rated 450 shp, driving Hartzell 93-in (2.36-m) diameter, three-blade, reverse-pitch propellers.

TBO
3,500 hours.

DHC6 Twin Otter

The all-can-do, super bush plane

Background

Many years before various post-war mergers afflicted the British aircraft industry (see page 196), de Havilland, the famous and internationally respected Hatfield-based concern, set up shop in Canada. Before the war they built 66 Moths of various kinds, then after the fighting started de Havilland Canada (DHC) turned out 25 DH82A Tiger Moth biplane trainers; 1,528 DH82C Tigers, modified to cope with the Canadian winter; 54 Fox Moths; and no fewer than 1,134 DH98 Mosquito fighter-bombers, surely the most potent warplane of any kind on either side in the war. When peace returned DHC designed a Tiger Moth replacement, the delightful Chipmunk, then the Canadian offshoot separated from its Hatfield parent and embarked on a policy of specialization. In this highly competitive world of ours a specialist will usually beat the Jack-of-all-trades and DHC have certainly become leaders in their field: the design and production of Short Take-off and Landing (STOL) aircraft. It started with the DHC2 Beaver and continued with the larger but still single-engine DHC3 Otter. Then came the DHC4 Caribou, a piston twin similar in size to the old Dakota. Currently DHC are marketing a fifty-seat, four-turboprop STOL airliner called the DASH 7 and a smaller twin turboprop, the Dash 8. The subject of this test report is another of their products that has enjoyed outstanding success all over the world, the DHC6 Twin Otter, which emerged in the mid 1960s. It will carry up to twenty-one passengers and cope with airfields that look almost like helipads.

Engineering and design features

In layout the Twin Otter looks like an overgrown Cessna 152 (see Chapter 2). The parallel-chord, high-aspect-ratio wing is braced to the fuselage by single struts and wide-span slotted flaps are augmented when lowered with Frise ailerons that droop, while still retaining their ability to move differentially. A PT6A-27 free-turbine engine (620 shp) is hung well forward of each wing at the point of strut attachment. Although wing-extension fuel tanks are available as an option, adding another 75 Imp/90 US gal (341 litres) capacity, standard fuel, ie, 307 Imp/370 US gal (1,400 litres), is carried in two banks of four fuel cells situated below the cabin floor. The front cells feed the right-hand engine and the rear four supply the left. Each set of four is fitted through a single filler point and fuel can be shunted from one set to the other when crossfeed is required to cope with an en-route engine failure. The leading edges of all flying surfaces may be fitted with de-icing boots. The massive swept fin requires no assistance

from additional dorsal or ventral areas. It carries a 20-ft span fixed tailplane several feet above the top of the fuselage. When the Twin Otter is used as a floatplane small additional fins are added to each half of the tailplane.

The fixed nosewheel undercarriage has powered steering and the mainwheels are carried on rigid tubular legs that provide shock absorbing by sliding up and down in slots let into the fuselage sides. Large, so-called 'high-flotation' wheels may be fitted for soft-field operations if required.

Six windows are set in each side of the cabin area, which is reached by a large double door behind the wing on the left-hand side of the shapely fuselage. There is a smaller door directly opposite on the right. Since the floor is about 5 ft above ground level – unusual for a high-wing mono-plane but necessary when the 400 Imp/480 US gal (1,817

De Havilland Aircraft of Canada DHC6 Twin Otter, a 20-passenger light transport with STOL capabilities.

litres) fire-fighting belly pack is fitted – steps are required to get in. Up front are two small doors for the crew, one on each side of the flight deck. Suitable steps and hand-holds are built into the fuselage but there is no door stay to keep it open and you enter at risk of having it blow into your face – a bad design oversight that. There are two baggage holds: one in the nose and another behind the passenger cabin, each with its own door. Total capacity is 800 lb (363 kg). An electro-hydraulic power pack operates the flaps, nosewheel steering, wheel brakes and, when fitted, ski retraction, so that these can be raised to allow a landing on wheels when flying between snow and tarmac.

Of course beauty is in the eye of the beholder but I found the Twin Otter a striking if slightly gawky-looking bird, with its enormous fin, rather odd undercarriage and

a collection of flaps and aileron brackets that make the wing look like one of those old-fashioned ironmongers' shops. Yet from some angles it has a graceful appeal.

Cabin and flight deck

The passenger cabin is 18 ft, 6 in (5.64 m) long from the front bulkhead to the one at the rear. Headroom is 59 in (150 cm) and maximum width measured at about seat level is 63 in (160 cm), tapering slightly towards the rear. Seating arrangements are variable, a typical commuter layout being six rows of three (two on the right and one on the left, separated by a wide aisle) with a double-seat opposite the door and a three-seat couch at the rear of the cabin. For its weight the Twin Otter has a spacious interior. Standards of interior trim vary according to the intended use but for regional airlines the manufacturers offer adequate, though not luxurious seating and decor with the accent on durability rather than gracious living. All but the rearmost passengers have a generous-size adjacent window.

The flight deck is entered through the two crew doors or you can use the door set into the front cabin bulkhead. Layout is rather like a military aircraft in presentation. The centre of the panel carries the avionics on the right and two columns of engine instruments on the left. There is an annunciator top centre and the glareshield slopes down sharply, left and right, to provide good forward visibility. Directly below the engine instruments is a neat fuel management panel with electric fuel selectors, boost pumps, emergency fuel pumps and fuel contents gauges calibrated in pounds.

Two flight panels may be provided for two-crew operation although the aircraft may legally be operated by one pilot. Control wheel arrangement is interesting. A single column emerges from the centre of the floor and then branches left and right like a huge letter 'Y'. At the end of each branch is a control yoke, the captain's having behind it a horizontal lever that may easily be raised and lowered by the left fingers while holding the control wheel. The lever works the powered nosewheel steering: up for turn right, level for straight ahead and down to turn left. The steering is accurate, light to the touch and very convenient – a far more sensible arrangement than those steering knobs that really need a three-handed pilot during the take-off and landing.

In the centre of the roof, dangling down towards you, are the engine controls: two power levers, two propeller levers and a pair of fuel condition levers. The power levers have twist grips to release the idle stops and allow selection of Beta/reverse-thrust modes. There is also an annoying flap lever that requires you to release a locking device with the thumb before it can be moved. Why is it the engineers always manage to inflict on pilots some monstrosity in an otherwise excellent flight deck? There is a bad pressurization control in the Piper Aerostar (see Chapter 5), poor nosewheel steering on the BAe Jetstream (see further on in this chapter) – I could name them by the score. Flap movement is slow and stately. There is a good flap-position indicator mounted vertically within the windscreen divider and the flap system is of the 'follow-up' type.

To the left of the engine management section is another roof-mounted panel for the battery master switch and engine starting. For some reason the generator switches are on another panel. There are test buttons for checking the fire-warning bell, which is loud enough to make you jump out of the aircraft.

I thought the flight deck was a little untidy but apart from the rather odd position of the engine and flap controls it is all convenient to use.

In the air

A single start switch indented to the central OFF position is moved left or right to wind up the related engine. You fire up in standard PT6 fashion and then unfeather the propellers when ready to move off. The brakes are first class and after a few moments the nosewheel steering becomes second nature. There is up to 60 degrees arc of movement left and right of centre. For a normal take-off, flap 10 degrees is used but STOL departures of the type I am about to describe entail selecting 20 degrees. Remembering the 64-kt V_{mca} and the 80-kt blue-line speed, you open up to maximum torque on the brakes. With the control yoke held well back the brakes are released, the machinery lurches forward and at some ridiculously low airspeed (around 50 kt) leaps off the ground. At that stage the nose should be lowered slightly encouraging rapid acceleration to around 80 kt, but a fully loaded Twin Otter will be at 50 ft above the runway in just under 1,200 ft on a windless day, which is even shorter than the take-off distance of a Piper Seneca light twin (see Chapter 5). With a moderate wind blowing at, say, 15 kt, take-off distance is about 800 ft – not bad for a twenty-passenger turboprop.

Like all propeller-driven, high-wing aircraft the Twin Otter tends to be a little noisy at high propeller rpm and this rapidly encourages you to reduce power for the climb. Best rate speed is 100 kt and a fully loaded example will go up at over 1,600 ft/min. Being essentially a short-distance STOL aircraft that would spend most of its time cruising at between 8,000 and 10,000 ft, the cabin is not pressurized. Speeds vary according to weight but the maximum cruise at 12,500 lb (5,670 kg) is 182 kt, 155 kt being adopted when range is important – not fast, but good for a low-powered, twenty-passenger STOL aircraft.

I measured these power-off stalling speeds at a time when the aircraft was within 1,000 lb (454 kg) of its maximum weight:

Flap setting	IAS
0°	74 kt
10°	64 kt
20°	60 kt
37½° (max)	58 kt

A slight tendency for the right wing to drop did not develop and recovery using the standard technique was almost instant. With power you can fly at some pretty low airspeeds. I got down to 45 kt without difficulty, although I would imagine that if a wing did drop following a low-speed stall what happened next would take some stopping. At 65–70 kt the Twin Otter may be thrown around the

OPPOSITE ABOVE: *The passenger cabin is basic but adequate for the short journeys intended for this light turboprop.*

OPPOSITE BELOW: *The various brackets under the wings are for the flaps and the drooping ailerons.*

sky with confidence. Visibility while turning is superb and handling is in the true de Havilland tradition – what could be better?

One-engine flying is perfectly straightforward, but the fully loaded single-engine climb of 340 ft/min is only just acceptable for a turboprop. The strange thing is that although the elevator and rudder trims were manual on the example flown, that for the ailerons was operated electrically.

It has always been a feature of de Havilland aircraft that they want to climb and fly (some other aircraft are kept in the air by brute force and pilot cunning). The Twin Otter is a typical de Havilland bird and 200 ft can be gained if the nose is not depressed while the flaps are lowered. Steep descents into jungle airstrips can be made by lowering full-flap and placing the propellers into Beta but *not* reverse pitch, otherwise it would be a case of 'Does anyone want to buy a tall, thin Twin Otter?'

For a normal short landing you approach at 75 kt, go for 65–70 kt over the threshold, land, then operate the power-lever twist grips to enter the reverse-pitch regime. With the aid of the powered brakes you can be landed and stopped from the 50-ft point in little more than 1,000 ft.

Capabilities

Maximum take-off weight is 12,500 lb (5,670 kg) and a typical commuter version would have an operating weight, including 2 pilots, of 7,320 lb (3,320 kg). That leaves 5,180 lb (2,350 kg) for passengers and fuel. Standard fuel weighs 2,457 lb (1,115 kg), leaving a payload of 2,723 lb (1,235 kg) – enough for 14 passengers and their baggage. That little lot could be flown for 600 nm at 182 kt or 720 nm at long-range speed. Twenty passengers and their baggage may be flown for 250 nm at 182 kt, all ranges quoted allowing reserve fuel for a 115-nm diversion and a 45-minute hold. So for its power (a pair of 620 shp engines is not all that much for a 20-passenger plane) the Twin Otter has above average payload/range. In terms of fuel economy, it manages 39 passenger nm/Imp gal, 32/US gal (8.57/litre). The landing weight restriction is no hardship in this class of aircraft. It entails not landing for about twenty minutes after take-of.

Verdict

The main competitor facing the Twin Otter is the Shorts Skyvan (see Shorts 330 test at the end of this chapter), which offers a similar number of seats and flies on a slightly more powerful pair of Garrett single-shaft engines. Although the Skyvan is used as a small passenger plane, its prime role is that of a freighter. Here it has an advantage over the Twin Otter in being able to accept more bulky consignments in its 6 ft, 6 in (1.98 m) wide, 6 ft, 6 in high cabin. The entire rear of the fuselage lets down to provide a convenient loading ramp.

As a commuter plane the Twin Otter has been more successful, selling in far greater numbers. You see them on floats, on skis (The British Antarctic Expedition used them

to great effect for a number of years) and as fire-fighters fitted with an under-fuselage water tank. The Shorts Skyvan is pleasant enough to fly but is a pig on the ground, and from the pilot's point of view the Twin Otter is less like hard work when frequent take-offs and landings are entailed. The Twin Otter is built to high standards of engineering and it has given excellent service on the regional airlines for many years.

Facts and figures

Dimensions

Wing span	65 ft
Wing area	420 sq ft
Length	51 ft, 9 in
Height	19 ft, 6 in

Weights & loadings

Max take-off	12,500 lb (5,670 kg)
Max landing	12,300 lb (5,579 kg)
Operating empty (ie, with crew)	7,320 lb (3,320 kg)
Useful load	5,180 lb (2,350 kg)
Seating capacity (with crew)	22
Max baggage	800 lb (363 kg)
Max fuel Standard	307 Imp/370 US gal (1,400 litres)
Optional	382 Imp/460 US gal (1,741 litres)
Wing loading	29.8 lb/sq ft
Power loading	10.08 lb/hp

Performance (standard fuel system)

High-speed cruise (10,000 ft)	182 kt
Long-range cruise (10,000 ft)	155 kt
Max range at 182 kt	600 nm
Max range at 155 kt	720 nm
Rate of climb Two engines	1,600 ft/min
One engine	340 ft/min
Take-off distance over 50 ft (STOL technique)	1,200 ft
Landing distance over 50 ft (STOL technique)	1,050 ft
Service ceiling Two engines	26,700 ft
One engine	11,600 ft

Engines
2 × Pratt & Whitney of Canada Ltd PT6A-27 free-turbine developing 620 shp, driving Hartzell 102-in (2.59-m) diameter, three-blade, reverse-pitch propellers.

TBO
3,500 hours.

Piper PA-42 Cheyenne III

The ultimate stretched Navajo

Background

For a short while Piper made a pressurized version of their Navajo piston-engine twin (see Chapter 5). For some reason it never caught on but in 1972 a turboprop version was devised which they called the Cheyenne. The original Navajo flew with engines of less than 300 hp; the PT-6 motors fitted to the turboprop version were flat-rated at 620 hp each. Not only was the Cheyenne double the power of the Navajo but, I suspect, its centre of gravity was too far aft, making it divergently unstable in pitch; raise the nose from its trimmed attitude, release the wheel, and the undulations that followed built up until you had to grab the controls and take over. The British Civil Aviation Authority never certified it in the UK.

Piper have since devised a little version called the Cheyenne I, a charming small turboprop in the same class as the Cessna Corsair (see airtest in this chapter). They have also gone the other way and developed a big offspring, the Cheyenne III, which was introduced in 1981. And if the original Piper turboprop is emblazoned on my mind as one of the worst aircraft I have flown, the Cheyenne III ranks among my happier moments in the air.

Engineering and design features

Although the outer wing panels are basically those of the Navajo, a totally new centre section, with pronounced sweep-back of the leading edges, has been designed. Fuel is carried in no fewer than ten tanks although fortunately there are only two filler points to each wing, which also has a large, leading-edge tank, an inboard fuel cell, a nacelle tank, an outboard cell and a streamlined wing-tip tank. In all, the system provides storage for 480 Imp/578 US gal (2,188 litres). The wing, tailplane and fin leading edges have pneumatic de-icing boots.

Electrically operated slotted flaps are provided with the same complicated electronic balancing system as installed on the Navajo, the pilot being able to adjust flap settings by as little as 2 degrees if required. The undercarriage has single wheels all round. It is raised and lowered by a hydraulic system powered by two pumps, one on each engine. There is an emergency hand pump, but in the unlikely event of complete hydraulic disaster – even loss of all fluids – each undercarriage leg has its own compressed nitrogen bottle.

The Cheyenne III is powered by a pair of Pratt &

Piper Cheyenne III, largest turboprop from this manufacturer. It owes its ancestry to the original Piper Navajo piston-engine light twin.

Whitney of Canada Ltd PT6A-41 free-turbine engines, which are flat-rated to 720 from their potential 850 shp. They drive three-blade propellers with 'Q' tips that are bent through 90 degrees to reduce erosion and increase ground clearance. At the same time, end losses are reduced and, in my experience, such propellers are noticeably quieter. They are provided with the usual Beta and reverse-pitch facilities, and to guard against accidental selection in the air the power levers are prevented from moving back behind their idle stops by a 'squat' device which automatically comes into play when the weight of the aircraft is removed from the undercarriage legs.

To minimize engine and propeller noise within the cabin the nacelles are set wide apart. Each one has an 8-ft long baggage locker that can take up to 200 lb (91 kg). Another 300 lb (136 kg) can go in the nose and a similar load may be placed at the rear of the passenger area so, in total, there is provision for 1,000 lb (454 kg) of luggage. At most, even when ten passengers are carried, half that amount would be more than they would be allowed travelling by commercial jet.

A very large fin, which extends forward to the rear of the cabin and is supplemented by a ventral area, ensures good directional stability and its big rudder was an obvious necessity to cope with possible failure of one of those powerful engines situated so far from the aircraft's centreline. Even so, V_{mca} is 97 kt, but in the Cheyenne III we are approaching the upper end of the light-aircraft league.

The flight deck is nicely planned. Below the glareshield is an annunciator strip, the engine readouts are to the right of the captain's flight panel and everything falls nicely to hand.

On top of the fin is a wide-span tailplane with separate elevators.

The long, lean fuselage is based upon Navajo frames. It has five rectangular cabin windows and, considering the 6.3 psi pressurization, they are of very large area. The flight deck has two even larger side windows and a big, two-piece windscreen.

Although the Cheyenne III is the product of typical Piper evolution – a bit of this and a bit of that added to a little something new – the overall visual effect is not too hard on the eyes. I would describe the aircraft as striking rather than pretty.

Cabin and flight deck

A large passenger door on the left-hand side lets down in one piece to provide three steps up to the rear of the cabin. On its right a second adjoining door may be hinged up to allow loading of bulky packages. The cabin is 53 in (135 cm) high, 51 in (130 cm) wide and at almost 23 ft from front to rear, probably the longest among light turboprops. Various internal arrangements are available. Apart from the two crew seats up front there can be a six-passenger layout with two folding tables, a refreshment centre and a toilet. Or you could have a high-density interior with eleven forward-facing seats (including the two on the flight deck). The ceiling and upper wall panels are of the sculptured, airliner type and very pleasing indirect lighting is provided along with the usual individual reading lights. A small but practical centre aisle allows movement from one end of the cabin to the other, and the interior is reminiscent of an opulent executive suite.

The flight deck is Piper's best. Most of the electric services along with engine starting are banished to a roof panel. A large area in the centre of the main panel accommodates the comprehensive avionics installation which includes a weather radar. Two flight panels are provided, and the engine instrument are in two vertical columns (one for each engine) reading torque, propeller rpm, ITT (turbine temperature), gas generator rpm (in percentages of maximum), fuel flow (in pounds per hour), and oil temperatures/oil pressure in one instrument. A comprehensive annunciator runs in a long strip built into the glareshield, electric circuit breakers are on each wall, cabin pressurization is handled low down to the right of the power pedestal, and nearby is the flap selector with its position indicator.

To the left of the pedestal is the undercarriage lever and its status lights, the parking brake handle and, below the two columns of engine readouts, a large single instrument with two pointers giving fuel quantities in the wings.

The comfortable seats adjust in most respects and, one way or another, the Cheyenne III flight deck is something of a pilot's paradise. A sea-level cabin can be maintained up to altitudes of 14,600 ft; at 28,500 ft the cabin is at only 8,000 ft and you can climb to 33,000 ft yet still experience 10,000 ft conditions inside the aircraft.

In the air

Starting follows standard practice for a PT6 engine, but whereas in many turboprops you must wait for the battery to recover before spooling up the second engine, an excellent feature of the Cheyenne III is its ability to shunt current from the running engine to the starter of the second one. For an aircraft weighing 11,000 lb (4,990 kg) the nosewheel steering is commendably light and it is easy to taxi using Beta mode to prevent excess speed on the ground.

Minimum control speed is 97 kt, V_r 105 kt, blue line 115 kt and cruise/climb speeds are recommended at 140 kt up to 18,000 ft, 130 kt from 18,000 to 24,000 ft and 115 kt above that altitude. These are, of course, indicated speeds. Although a rotate speed of 105 kt is rather higher than your usual run of two-seat trainers, taking off in a Cheyenne III makes few demands on the pilot except that he must not exceed 1,895 ft:lb torque. We were anxious to depart the London TMA without delay and continue climbing under radar control so I used 160 kt initially. The VSI said we were going up at 2,000 ft/min but if you adopt the speed for best rate of climb it will hit 2,500 ft/min at high average weights.

At 25,000 ft maximum cruise power for the conditions (ISA + 10°C) entailed setting 1,900 rpm on the propellers and 1,450 ft:lb torque. This gave us a TAS of 280 kt for a total fuel burn of 570 lb/hr. The manual also quotes figures for 'Recommended cruise power' and 'Long-range cruise power', the latter being based on 1,700 rpm for the propellers. We used 1,090 ft:lb of torque to produce a TAS of 228 kt for a total fuel burn of 425 lb/hr.

In all phases of flight the Cheyenne III is one of the quietest turboprops I have flown. People in the cabin can hear those on the flight deck and you would have to move up in class to fanjets for more silence.

The primary controls are first class, a pleasure to handle and beautifully harmonized. I thought the elevator and aileron trimmers worked well but when later I shut down one engine the rudder trim was not man enough to maintain balanced flight without assistance on the rudder pedals. However, the example flown was one of the earliest demonstration models rushed over for the Farnborough Air Show and I understand the problem has since been dealt with. There is slight, if not very pronounced, lateral stability and excellent pitch and yaw damping. Frankly I have never been a fan of Piper handling but they have done their homework on the Cheyenne III.

Without warning the Piper demonstration pilot sitting on my right shut down the right-hand engine. We were at 15,000 ft at the time and without adding power on the live engine there was a not-to-be-sneezed-at TAS of 180 kt. Single-engine ceiling at average weights is around 17,000 ft and there is a maximum-weight, one-engine climb of 565 ft/min – real twin-engine safety if ever there was!

Initial approach was flown at 130 kt aiming for 111 kt on short finals. During the round-out the power levers are brought back to the stops and after landing the nosewheel is lowered before applying reverse thrust, which shortens the landing roll by about 400 ft.

Capabilities

An average-equipped Cheyenne III would have an empty weight of around 6,630 lb (3,007 kg), leaving a useful load of 4,655 lb (2,122 kg) of which 2,720 lb (1,234 kg) may be placed in the cabin. With full tanks, ie, 3,886 lb (1,763 kg), payload is 769 lb (349 kg) – enough for a pilot and three passengers, each carrying the usual 44-lb (20-kg) baggage allowance. Range is then 1,800 nm at maximum cruise power, or 2,200 nm using the long-range cruise technique but, with respect to most turboprops, they are not really fast enough for this kind of distance. A more typical mission for a Cheyenne III would be a pilot and seven passengers plus 44 lb (20 kg) baggage each, totalling about 1,650 lb (748 kg), and allowing 3,000 lb (1,361 kg) of fuel to be carried which would give it a range of more than 1,000 nm at a cruising speed of 287 kt. At this speed, which takes it out of the piston-engine class, passenger nautical miles per Imperial gallon work out at 23.15 (assuming seven passengers) and 19.29 US gal (5.09 nm/litre). Turboprops use more fuel than piston engines, but this level of economy for a fast executive aircraft is satisfactory.

Verdict

Nearest competitor to the Cheyenne III is the longer-established Beech King Air 200, which is slightly larger (see next airtest) and uses the same engines, but at 850 shp, not flat-rated to 720 shp as in the Piper offering. Being that little bit larger, the King Air 200 will carry another four people in the high-density layout. Here are a few comparative figures for the two aircraft.

	Piper Cheyenne III	Beech King Air B200
Useful load (stndrd)	4,896 lb (2,221 kg)	5,052 lb (2,292 kg)
High-speed cruise	280 kt	280 kt
Range (high speed)	1,800 nm	1,757 nm
Rate of climb	2,400 ft/min	2,450 ft/min
Cabin height	53 in (135 cm)	57 in (145 cm)
Cabin width	51 in (130 cm)	54 in (137 cm)
Cabin length	22 ft, 11 in	16 ft, 8 in

In the air the Cheyenne III's widely spaced engines are a striking feature.

I thought the Cheyenne III was quieter and more pleasant to fly than the King Air 200 but prospective buyers would have to study these two fine aircraft in detail and relate them to their requirements before reaching a firm decision. The King Air 200 may be landed at its 12,500-lb (5,670-kg) maximum take-off weight whereas the Piper offering must lose 870 lb (394 kg) when at maximum weight before a landing may be made. Some would argue

that the most likely time a Cheyenne III would take off at its maximum 11,200 lb (5,080 kg) would be when a long journey is planned, but 870 lb represents 75 minutes' fuel, which is a long time to wait in the air if, for some reason, a return home is necessary. The Beech aircraft is built to higher engineering standards and this is reflected in its price. But the Cheyenne III is a fine piece of aeronautical hardware and a lot of aircraft for the money.

Facts and figures

Dimensions

Wing span	47 ft, 8 in
Wing area	293 sq ft
Length	43 ft, 4 in
Height	14 ft, 9 in

Weights & loadings

Max ramp	11,285 lb (5,119 kg)
Max take-off	11,200 lb (5,080 kg)
Max landing	10,330 lb (4,686 kg)
Max zero fuel	9,350 lb (4,241 kg)
Equipped empty	6,630 lb (3,007 kg)
Useful load	4,665 lb (2,112 kg)
Seating capacity (with crew)	9/11
Max baggage	1,000 lb (454 kg)
Max fuel	480 Imp/578 US gal (2,188 litres)
Wing loading	38.22 lb/sq ft
Power loading	7.78 lb/hp

Performance

Max cruise speed at 20,000 ft	290 kt
Long-range cruise speed at 33,000 ft	225 kt
Range at max cruise speed (20,000 ft)	1,400 nm
Range at long-range cruise speed (33,000 ft)	2,200 nm
Rate of climb Two engines	2,400 ft/min
One engine	565 ft/min
Take-off distance over 50 ft	3,100 ft
Landing distance over 50 ft (with reverse thrust)	2,300 ft
Service ceiling Two engines	33,000 ft
One engine	17,700 ft

Engines
2 × Pratt & Whitney of Canada Ltd PT6A-41 free-turbine developing a flat-rated 720 shp, driving Hartzell 95-in (2.41-m) diameter, three-blade, reverse-pitch propellers.

TBO
3,000 hours.

Beechcraft Super King Air B200

Top end of the best-selling turboprop range

Background

Popular among business operators in the 1960s was the Beech Queen Air which was offered with a choice of 340- or 380-hp piston engines. In many respects they were excellent and practical light twins, though inclined towards the short and dumpy. Also, if you dared to mention the name Queen Air during a routine airways call some comedian was bound to press the button and scream 'Hello sailor!'

From the Queen Air, Beech developed their King Air series and in the process became the first manufacturer to enter the light turboprop market. The King Air range has become the best-seller of its class. Lowest powered is the King Air C90 which flies on two 550-shp PT6A-20 engines. Other models include the F90 with another 200 hp a side and cruising speeds of almost 270 kt. Top model is the B200 which has been inflicted with the name Super King Air, one that seems to cheapen what is in most respects a very fine, high-quality turboprop. Beech have deservedly collected a reputation for above-average standards of engineering. Assessed within the general aviation context Beech aircraft are as good as you can get. At one time the model C90 was top-seller among the King Air fraternity but the 200 series is about to exceed it in number of units sold, which is remarkable because it has not been around for as many years as little brother.

Engineering and design features

To provide the additional area needed for the much larger and heavier model 200, the original King Air wings have extended tips and a much wider centre section which holds the engines well clear of the fuselage sides. While this is conducive to a quiet cabin it brings with it greater yawing moments during asymmetric flight. These are countered by providing a large fin and a rudder with automatic bias powered by a two-sided piston which is supplied with pressure from the two engines. When one motor fails, pressure is no longer balanced and the piston moves over, applying corrective rudder. The idea was originally pioneered by the de Havilland concern when they built the first 125 business jets.

Slotted flaps in four sections are moved electrically and controlled from the flight deck on a lever gated at the approach setting (40 per cent maximum). Fuel is carried in two separate systems, one in each wing. There is a nacelle tank, two wing leading-edge tanks, two interspar

This top-of-the-range turboprop is slightly larger than the Piper Cheyenne III, which is its nearest competitor.

or box tanks (in the form of bladders) and a large integral or wet-wing tank. Each side of the wing centre-section holds an auxiliary tank and the entire system has a total usable capacity of 450 Imp/544 US gal (2,059 litres). It sounds complicated but there are only two filler points on each side of the aircraft. Pneumatic de-icing boots are fitted to the leading edges of the wing, fin and tailplane.

PT6A-42 free-turbine engines, which deliver 850 shp, replace the earlier -41 motors. There is no increase in power but the -42 engines have improved performance at high altitudes and the B200 King Air will cruise at up to 35,000 ft (4,000 ft higher than the earlier model), where it can mix it with the jets. Consequently, pressurization differential has been increased by 0.5 psi to ensure a 10,380-ft cabin at the new maximum altitude. Personally I would regard an 8,000-ft cabin as plenty high enough for the average past-middle-aged company director; however, the B200 could provide these conditions while flying at a still useful 30,000 ft. The engines are slung well forward of the wings on long nacelles.

The undercarriage, which has double mainwheels, each with its own disc brake, retracts in only four seconds. Although the fuselage is based upon the old Queen Air, it is now stretched beyond recognition. Also, the original, square cabin windows have been replaced by large circular portholes, seven on each side. There are dorsal and ventral fins and the tailplane sits 'T' fashion on top of the main vertical area. The wing is set well forward along the fuse-lage. On the left side, behind the wing, is a large, one-

piece door that lets down to reveal five in-built steps. It needs them because the 200 stands quite high off the ground.

The King Air 200 is handsome and reeks of good quality but is not beautiful to look at. Few turboprops are.

Cabin and flight deck

The passenger area may be fitted out as a six- or eight-seat executive suite with all the usual amenities (tables, refreshment centres, separate toilet, and so forth), or twelve forward-facing seats can be installed. Conversion from one interior to the other takes less than one hour. The cabin is 57 in (145 cm) high, 54 in (137 cm) wide and 16 ft, 8 in (5.08 m) long from behind the flight deck – shorter than some in its class, but adequate. The windows have polaroid inner screens which may be rotated to reduce or completely eliminate the glare of the sun. Behind a partition at the back of the cabin is the toilet followed by the baggage hold which has a capacity of 410 lb (181 kg) – enough for nine passengers but, in my view, no more; the King Air could do with a little more baggage capacity for an aircraft of its size. Type of upholstery and colour scheme is at the discretion of the purchaser but Beech turn out a nice aircraft, inside and out.

A sliding door at the front of the cabin leads to the flight deck. Fuel management, related circuit breakers, and so on are on a panel under the left-hand window. The various internal and external lights and the windscreen wipers are

OPPOSITE: Various interiors are available. This is a six-seat corporate aircraft layout with folding tables, but up to twelve seats may be fitted in the high-density configuration.

Although the Super King Air is based upon an older design, its high quality and excellent performance have made it one of the best-selling light turboprops.

handled on a small roof panel.

The glareshield carries a pair of caution lights for each pilot. In effect they tell you to consult the annunciators of which there are two: one in the glareshield for the nasties (engine fire, lost fuel pressure, and so forth), and a larger one bottom centre that gives the status of various systems. The radio installation is nicely arranged with the weather radar in the centre surrounded by all the other avionics. To the right of this large area is the copilot's flight panel and to the left are the engine readouts in two columns, one for each motor. From top to bottom they run in the following order: interstage turbine temperature, torque (in ft:lb), propeller rpm (in hundreds and thousands like a piston engine), gas generator rpm (in percentages of maximum), fuel flow (lb/hr) and combined oil pressure/oil temperature. Below the captain's flight panel are the electric switches and the copilot has a matching panel devoted to cabin temperature etc. A large console drops down from the centre of the panel and extends back between the pilots. At its upper end are some pressurization instruments followed by the engine levers (power, propeller and fuel conditions). There is a flap control, the three trimmers and then the pressurization controls. It is a pity they have not grouped all pressurization together. Finally, at the back of the console is the autopilot control box. Although not the most modern flight deck, there is not much to complain of in the King Air and it is a pleasant working area for the pilots, the seats providing good support for the long trips that are within the capabilities of this aircraft.

In the air

Despite its size, and the King Air B200 is at the top limit for single-crew operation, it manages nicely without powered nosewheel steering. Turning radius is only 19 ft, 6 in when measured at the nosewheel. For the flight we had 2,300 lb (1,043 kg) of fuel and seven of us on board. There was enough payload left to add another 1,000 lb (454 kg) of fuel or fill all seats and take on maximum baggage. V_{mca} is 90 kt, V_2 came at 105 kt and blue-line (V_{yse}) was marked on the ASI at 125 kt. Maximum torque during take-off is 2,230 ft: lb and no flap is used. Acceleration is rapid, there is not the slightest tendency to swing and very soon we were at our recommended 160-kt climbing speed with the VSI reading more than 2,400 ft/min. Later I timed from 9,000 ft to 11,000 ft in exactly 60 seconds.

At Flight Level 250 (25,000 ft altitude on a standard day) 'Maximum cruise power', which entailed setting 1,800 rpm and 1,700 ft:lb torque, gave us a TAS of just under 290 kt, which is almost 15 kt faster than the previous model and in a different league to the piston-twins. 'Normal cruise power' (these are the terms used in the Beech flight manual, by the way) uses the same propeller rpm but torque is reduced to 1,696 ft:lb and the TAS on a standard day is 284 kt for a total fuel burn of about 600 lb/hr (272 kg/hr), or 75 Imp/89 US gal (337 litres) per hour. If my memory is not at fault the aircraft is a little noisier than the smaller King Air C90. And while not as quiet as the Piper Cheyenne III it is never unpleasant, not even during the take-off.

The aircraft handles nicely with particularly fine ailerons but it could do with a little more stability. Surprisingly the C90, with its shorter fuselage, has more stability in pitch. There is a tendency among modern designers to rely heavily on the autopilot but I firmly believe that the basic aircraft should have good, in-built corrective stability in all three axes. All this is of academic interest to those riding in the cabin and while I would not wish to give the impression that the King Air B200 is unstable, for its size it could be better. Certainly it is a lively bird; even at 29–31,000 ft the rate of climb was around 800 ft/min and at lower levels it can almost equal that on one engine. Asymmetric handling is impeccable.

The clean stall came at 90 kt and maximum flap reduced this by about 20 kt, in each case with a slight tendency for the right wing to drop. When joining the circuit the wheels can be lowered at a useful 182 kt and approach flap (40 per cent) even sooner (200 kt), so there is plenty of drag available for slowing down. Initial approach is flown at 125 kt followed by 95 kt as the fence is crossed. After touchdown the nosewheel was lowered, the power levers lifted and then brought back into reverse thrust, which can reduce the landing roll by 600–700 ft.

Capabilities

A well-equipped King Air B200 would have an empty weight of 7,950 lb (3,606 kg), leaving a useful load of 4,640 lb (2,105 kg), of which 3,050 lb (1,384 kg) may be placed in the cabin (that is enough for fifteen people and their luggage). Maximum fuel weighs 3,644 lb (1,653 kg), leaving 996 lb (452 kg) for payload – a pilot and four passengers plus their luggage, for example. Full tanks will fly a with-reserve 1,766 nm at 280 kt ('Maximum cruise power' at 31,000 ft), or almost 2,000 nm at 271 kt using 'Maximum range power'. Alternatively, eight people (pilot and seven passengers) could fly for more than 1,500 nm at 285 kt, or a pilot plus eleven passengers would be able to do a nonstop, 1,050 nm. A full aircraft (for instance, two pilots and thirteen passengers with their baggage) could be flown for over 700 nm. The King Air 200 improves on the Piper Cheyenne III passenger miles per unit of fuel by about 14 per cent (see previous airtest).

Verdict

The King Air B200 is one of the fastest corporate turboprops on the market. It is also the biggest in general aviation terms. For journeys of up to 500 nm or so it is not a lot slower than the Cessna Citation II light fanjet (described in the next chapter) but on longer trips even a relatively slow jet can save time. For example, on a 1,000-nm journey the Citation II will get there 48 minutes before the King Air B200 and give its passengers a quieter ride. The two aircraft cost about the same to buy but the King Air burns considerably less fuel and can offer up to another five seats.

When compared with Piper's biggest turboprop, the Cheyenne III, the Beech is more powerful, can provide another four seats and is probably a better piece of engineering. This is, of course, reflected in the price, the King Air B200 being considerably more expensive than the Cheyenne III. It can also be landed at its maximum take-off weight, which under some circumstances could be an advantage. Of course, when evaluating aircraft at this level, comparisons are between one piece of excellence and another. And judging by the way King Airs continue to sell they must be pretty excellent.

Facts and figures

Dimensions

Wing span	54 ft, 6 in
Wing area	303 sq ft
Length	43 ft, 9in
Height	15 ft
Cabin height	57 in (145 cm)
Width	54 in (137 cm
Length (ex flight deck)	16 ft, 8 in

Weights & loadings

Max ramp	12,590 lb (5,711 kg)
Max take-off & loading	12,500 lb (5,670 kg)
Max zero fuel	11,000 lb (4,990 kg)
Equipped empty	7,950 lb (3,606 kg)
Useful load	4,640 lb (2,105 kg)
Seating capacity (with crew)	14
Max baggage	410 lb (186 kg)
Max fuel (with auxiliary tanks)	450 Imp/544 US gal (2,059 litres)
Wing loading	41.25 lb/sq ft
Power loading	7.35 lb/shp

Performance

Max speed	294 kt
Max cruise at 31,000 ft	280 kt
Normal cruise at 31,000 ft	271 kt
Range at max speed (31,000 ft)	1,766 nm
Range at normal cruise (31,000 ft)	1,974 nm
Rate of climb Two engines	2,450 ft/min
One engine	740 ft/min
Take-off distance over 50 ft	3,345 ft
Landing distance over 50 ft (with reverse)	2,074 ft
Service ceiling Two engines	

Engines
2 × Pratt & Whitney of Canada Ltd PT6A-42 free turbines developing 850 shp, driving 98½-in (2.5-m) diameter, three-blade, reverse-pitch propellers.

TBO
3,000 hours.

Embraer EMB-110 P2 Bandeirante

The Brazilian bombshell

Background

The Embraer outfit, or to give its full, unabbreviated title, Empresa Brasileira de Aeronautica, is a newcomer among aircraft manufacturers. Until its formation the only aircraft being built in Brazil were a few locally designed trainers. It so happens that adjacent to Sao Jose dos Campos Airport (not all that far from Sao Paulo) is an air-force technical school and the Research and Development Institute. One day in 1965 the local students were told: 'Go and design a light turboprop capable of flying sixteen passengers in reasonable comfort.' Project IPD 6504, the name given to this paper aircraft, was never intended as anything more than an exercise for the students. But when they had done their bit, people started to jump up and down and there were cries of: 'We must build this come hell and high water.' (In Brazilian, of course.) So the government sat up, took notice and acted. A new factory was built at the airport, staff were sent for training, Piper light planes were built under licence to gain experience and gradually the turboprop designed at the nearby technical school took shape.

They put in charge one Ozires Silva, an air-force colonel who once told me that he had no previous commercial experience. Yet within a decade Embraer were building ag-planes of their own design; an Italian jet fighter under licence; the Piper Cherokee, Arrow and Navajo; the Tucano turboprop light ground-attack/military trainer; the Xingu pressurized business turboprop (both of them Embraer designs); and Project IPD 6504, which by now had collected the name Bandeirante. In addition, a thirty-seat turboprop commuter plane was being developed and will have flown by the time you are reading this.

Embraer was formed in January, 1970, it was making profits after five years, and now it has become the fifth largest general aviation manufacturer in the world. The subject of this airtest is their Bandeirante, by now stretched to carry eighteen or nineteen passengers. Why did I call it the 'Brazilian bombshell' in my subtitle? Because it has hit the world aviation market like a tornado and outsold all comers in its class. By mid-1982 more than 450 of them had been delivered world wide.

Engineering and design features

The airframe is an utterly conventional riveted light-alloy structure with a moderately tapered wing spanning just over 50 ft. Electrically operated double-slotted flaps are controlled on an infinitely variable selector with a position

Embraer EMB-110 P2 Bandeirante, a light passenger airliner built to air transport standards.

indicator marked in percentages of maximum deflection. Two integral tanks in each wing make up the fuel system, which has a usable capacity of 366 Imp/440 US gal (1,665 litres). The system is filled through the two outer tanks which feed into the inners. Quick-drain cocks are located in the wing roots for defuelling. The wing is located about half way down the fuselage, an arrangement dictated by the light weight of turboprop engines. De-icing is by pneumatic boots on the flying surface leading edges.

The two PT6A-34 free-turbine engines are carried well forward of the wing leading edges. They develop 750 shp and drive three-blade, reverse-pitch propellers. All legs of the undercarriage have single wheels, power-assisted hydraulic brakes are provided and the hydraulic-powered nosewheel is steered through a small control knob.

There is a swept fin and rudder, supplemented by small ventral and dorsal areas and the 25-ft span tailplane is one of the biggest I have seen on an aircraft of this size. Trim tabs occupy about half the span of the elevators. The Bandeirante is marketed in three versions: cargo, mixed cargo/passenger and all passenger. This report deals with the passenger version. The fuselage is about 48 ft long

The flight deck is easy to live with. Nosewheel steering is controlled on the circular handle near the captain's left knee.

with nine cabin windows on each side. There are two large entrance doors, one at each end of the cabin and these let down to near ground level. Four built-in stairs are provided along with a quite substantial handrail.

For its weight – 12,500 lb (5,670 kg) – the Bandeirante is a big aircraft, a giant among light planes.

Cabin and flight deck
The passenger cabin is 63 in (160 cm) wide, 63 in (160 cm) high and over 31 ft (9.4 m) long. At the rear is a separate baggage hold with a capacity of 529 lb (240 kg). There is also a small toilet. Internal arrangements can be varied with up to seven rows of seats arranged two on the right and one on the left with an aisle offset to the left. A more comfortable passenger load would be eighteen (six rows of three). The seats in the example evaluated were covered in high-quality, golden-brown leather, a material the Brazilians handle with great skill. All the usual air vents and reading lights are provided in this large cabin

which is very much in the small airliner tradition.

The flight deck is separated from the passenger area by a bulkhead. Seven windows surround the pilots and these provide good visibility. They are stressed to give protection from bird strikes. There are two sets of flight instruments and the central area of the main panel is devoted to radio. There is a small annunciator and an excellent fuel management area laid out in pictorial form. The powered nosewheel steering knob is on the left ledge where it falls conveniently to the captain's hand. A master caution light is built into the centre of the glareshield, alerting the crew to a malfunction which is then identified on the annunciator panel. The engines are controlled on a small pedestal which also houses the elevator, aileron and rudder trimmers. All the usual instruments required for turboprops are in two vertical columns to the right of the captain's flight panel. Because of the general spaciousness of this flight deck it is totally uncluttered.

In the air

Engine starting follows standard PT6 practice and some pilots prefer to taxi on one engine only to avoid excessive speed. Alternatively you can lift the power levers and engage the Beta mode. Yet another technique is to feather the propellers and taxi on the residual thrust from the four exhaust pipes. The Bandeirante will do that on level ground and it is quieter than using Beta. Nosewheel steering is excellent using the control knob with the left hand. It is fitted with a disengage button and during take-off this is pressed when the rudder becomes effective, at around 40 kt. V_{mca} is 84 kt, V_r at high average weights is 95 kt and V_2 (take-off safety speed) is 104 kt with blue-line marked on the ASI at 115 kt. The torque meters are red-lined at 1,970 ft: lb and on no account may this be exceeded. I lined up and advanced the power levers with my right hand while steering with my left. At 40 kt the disengage button was pressed, the left hand was transferred to the control yoke and within seconds it was time to lift the nose and take off. After V_2 I raised the undercarriage and the 25 per cent flap setting used for the take-off then reduced propeller rpm to 91 per cent. Rate of climb was in excess of 1,800 ft/min – we were 900 lb (408 kg) below maximum weight at the time.

At 5,500 ft the economic cruise settled at 191 kt TAS. Maximum cruise for the Bandeirante at its highest weight is 221 kt when flown at 10,000 ft, which is as high as this unpressurized aircraft is likely to operate while carrying passengers. In turboprop terms the noise level is average; there is room for improvement. An interesting feature of the Bandeirante is that weight does not materially affect cruising speed. The aircraft is very stable in all three axes and handling is superb with controls harmonized in the classic mould. It stalls like a flying-club trainer, only a little faster; wheels and flaps-up I recorded 90 kt and 71 kt when 100 per cent flaps are used. There is pronounced pre-stall buffet, when the nose bows gently down while the wings remain level.

An engine was 'failed' on me without warning and very little rudder was required to maintain direction against the live motor. If the mood takes you, direction can be maintained on the excellent ailerons alone, although to regain balanced flight the rudder must obviously be used. Fully loaded a Bandeirante will go up at more than 400 ft/min on one engine, which has got to be good.

An approach at high average weights should be made at 100 kt, going for 90 kt over the fence, but these figures reduce by up to 6 kt as landing weight drops to 10,000 lb (4,536 kg) and below. When the flaps come down there are hardly any trim changes. The aircraft is pleasant to land and after the nosewheel has been planted on the runway reverse thrust may be selected along with the powerful brakes when stopping distance is critical. Otherwise the brakes on their own will prove adequate on all reasonable-length runways. When the speed approaches 40 kt the nosewheel power button should be engaged so that direction can be controlled on the steering knob.

Commuter pilots are often called upon to make several take-offs and landings per hour so it is very important that aircraft operated on short routes should be easy to handle in the air and, equally vital, on the ground. At the end of a busy day the Bandeirante will return its happy pilots in a fit state of mind to enjoy their baked beans on toast.

Capabilities

Maximum ramp weight at 12,600 lb (5,715 kg) allows a useful load of 4,800 lb (2,177 kg) in an average equipped aircraft. The Bandeirante is essentially a short-range commuter plane and it will fly eighteen passengers at 214 kt over 215 nm with airline reserves. Alternatively fourteen passengers could be transported on a 500-nm sector or ten could be carried for 900 nm. These distances allow for starting, taxiing, take-off, climb, a 55-nm diversion and a 45-minute hold followed by the descent and landing. Maximum payload may be flown for 160 nm.

Maximum landing weight entails burning off about 480 lb (218 kg) of fuel, so for short journeys the highest take-off weight is likely to be around 12,250 lb (5,557 kg). The maximum zero fuel weight is 12,015 lb (5,450 kg), limiting the cabin load to 4,215 lb (1,912 kg), which is more than enough for twenty-one passengers and their baggage. A fully loaded Bandeirante is happy operating from 3,000-ft runways, provided they are not located at high altitude.

Operating economics are a big subject and too complex to discuss here. But the Bandeirante in its various forms is giving excellent service all over the world and its ability to turn in a profit has no doubt contributed to the success of the type. Being able to carry a lot of passengers at over 220 kt on relatively low power, the Bandeirante is a very fuel-efficient light transport, offering 46.8 seat nm/Imp gal when eighteen fare-paying customers are present and on parade, (comparable seat per mile figures for US gallons and litres are 39 nm and 10.29 nm, respectively). These figures are about 20 per cent better than the DHC6 Twin Otter described earlier in this chapter.

Verdict

The Bandeirante has been built for utilizations of around 2,000 hours per year. Even the best general aviation equipment would find the going hard when flown at that rate. So the airframe and its system are to a higher specification

The Bandeirante is a big aircraft for its 12,500 lb weight.

than you would normally find on a business turboprop which, in any case, would be unlikely to fly more than 600–700 hours per annum, probably less.

Beech used to market a slightly smaller unpressurized turboprop, the model 99, but it went out of production. Then along came the Bandeirante offering a few more seats, a bigger cabin and better economics. They were purchased by regional airlines all over the world, many of

them in the United States, where they can be seen flying commuters as well as freight on short routes. Now Beech are introducing a new commuter in a late-hour attempt to cash in on the commuter boom. In the United States the Bandeirante can be operated in slightly modified form under FAR-41 regulations. Maximum weight is increased by about 500 lb (227 kg) and there is a 420-lb (191-kg) gain in useful load, although some of this is lost by having

to fly the aircraft on a two-crew basis, since it now exceeds the 12,500-lb (5,700-kg) limit for one-pilot operations.

When you consider the few years Embraer has been in existence, it is the more remarkable that they have grown to their present size, with such a diverse range of aircraft. Partly this is due to good management leading an enthusiastic workforce. But the incredible success of the company could never have been realized without the world-beating Bandeirante.

Facts and figures

Dimensions

Wing span	50 ft, 3 in
Wing area	312 sq ft
Length	49 ft, 6 in
Height	15 ft, 6 in

Weights & loadings

Max ramp	12,600 lb (5,715 kg)
Max take-off	12,500 lb (5,670 kg)
Max landing	12,015 lb (5,450 kg)
Max zero fuel	12,015 lb (5,450 kg)
Equipped empty	7,800 lb (3,538 kg)
Useful load	4,800 lb (2,177 kg)
Seating capacity	23
Max baggage	529 lb (240 kg)
Max fuel	366 Imp/440 US gal (1,665 litres)
Wing loading	40.06 lb/sq ft
Power loading	8.33 lb/shp

Performance

Max cruise at 10,000 ft	222 kt
Long-range cruise at 10,000 ft	176 kt
Max range at 222 kt (10,000 ft)	900 nm
Max range at 176 kt (10,000 ft)	1,025 nm
Rate of climb	
Two engines	1,790 ft/min
One engine	430 ft/min
Take-off distance over 50 ft	2,300 ft
Landing distance over 50 ft	2,430 ft
Service ceiling	
Two engines	22,500 ft
One engine	11,000 ft

Engines
2 × Pratt & Whitney of Canada Ltd PT6A-34 free-turbine developing 750 shp, driving Hartzell 93-in (2.36-m) diameter, three-blade, reverse-pitch propellers.

TBO
3,500 hours.

British Aerospace Jetstream 31

The walk-tall commuter plane

Background

Most nations need protecting from their politicians; the British certainly do. After the Second World War the boys and girls of Westminster developed a lust for involving themselves in commerce. And one of their favourite pastimes was arranging shotgun marriages between the old-established aircraft manufacturers. Thus Britain lost such famous names as de Havilland, Vickers, Avro and Bristol.

One outfit that refused to be bullied into the merger game was Handley Page, famous in the First World War for its massive 0/400 biplane bombers which were the most successful on either side of the conflict. During the Second World War Handley Page built the Halifax, another successful heavy bomber, and when jet engines became available they devised the Victor, which in its day was the most advanced bomber in the world. It had always been a family enterprise with Sir Frederick Handley Page firmly at the head of the table. And Sir Fred would have none of this merger nonsense.

When it became clear that this magnificent company was being ostracized by the British Parliament, Handley Page decided to embark on a civil project: an eighteen-passenger twin turboprop which, at the time, was aimed at the corporate market. They called it the Jetstream and it emerged overweight, down on performance and plagued by a pair of brilliant but underdeveloped engines. Things got bad at Handley Page and Sir Fred opened discussions with the Hawker Siddeley Group (one of the government-inspired mergers), but on 21 April, 1962 the grand old man died and his successors on the board decided to continue the fight.

Soon afterward various hoped-for orders were cancelled, prospective customers got cold feet and Handley Page, by now overcommitted in engines, equipment and materials, became insolvent. It was the end of the road for this great company.

So it might have been for the Jetstream, but not long after the demise of Handley Page it was decided that the British Royal Navy and Royal Air Force needed some twin-engine turboprops. Jetstream wings had always been manufactured by Scottish Aviation at Prestwick Airport near Glasgow and they received a contract to set up a small production line and supply the armed forces.

Over in the US of A, a few of the regional operators who had bought the original Jetstreams decided to fit them with Garrett engines, which have been giving excellent service ever since. The idea caught on and it was not long before the Prestwick firm, by now the Scottish Division of

British Aerospace (an even bigger merger job but, thankfully, no longer nationalized) got the OK to develop a spruced-up civil version to be known as the Jetstream 31. I was fortunately able to fly one of the first demonstration models just before it left for a lengthy sales tour. The trip was just in time for this book.

Engineering and design features

The Jetstream 31 has been built to withstand utilizations in the region of 2,500 hours per annum, consequently it is built to the same engineering standards as a large transport aircraft. Walking round the factory at Prestwick you would have to be a mechanical moron not to recognize that this is no general aviation hardware. The wing has two spars running from root (where they bolt onto heavy fuselage frames) to tip. A third spar, which continues through the fuselage, extends for about two-thirds of the span.

Heavy-gauge top and bottom wing skins have spanwise stringers Redux bonded to them and two integral tanks (wet wings) have a total fuel capacity of 385 Imp/465 US gal (1,760 litres). The two Frise ailerons each have a servo tab to lighten control loads, and double slotted flaps may be set to UP, 10, 20 and 50 degrees via a 'follow-up' selector. After landing, lift-dump may be selected to bring the flaps down another 20 degrees.

The beautifully proportioned fin carries a fixed tailplane about half-way between the top and its lowest extent. If you must aim to keep the tailplane and elevators out of the slipstream area this arrangement looks better and is probably a superior engineering solution than those 'T' tails which seem to have captured the imagination of so many modern designers.

The Jetstream 31 is powered by Garrett TPE 331-10 engines which deliver 900 shp to four-blade, 106-in (2.69-m) diameter Dowty-Rotol propellers which rotate at only 1,590 rpm. These are very up-market and not to be confused with the kind of windshovel that forms part of the kit on even the better general aviation aircraft. The blades have an advanced airfoil section.

Garrett engines are of the single-shaft type which is confined to one rotating assembly, there being no separate gas generator and power turbine/gearbox group as is the case with PT6 and Allison turboprops. From idle to maximum power engine rpm remain constant. Movement of the power levers adjusts the amount of fuel being supplied to the burners and the propeller changes pitch to match the power being developed. A pilot's first landing

196

with this type of engine is an experience to remember. You round-out, pull back the power levers to the idle stops and the engine note remains unchanged, giving the impression that you and your favourite plane are about to fly through the far hedge. In the Jetstream these engines are mounted with the air intakes above the propellers where there is less risk of debris ingestion while on the ground. Overhaul life is a remarkable 6,000 hours or they can be operated using an 'on-condition' maintenance scheme.

The fuselage is clad in chemically etched skins and, like the remainder of the airframe, is subjected to the most rigorous anti-corrosive treatments. Riveted parts are wet-assembled to prevent airframe rot at the metal interfaces. Seven large oval windows are let into the sides of the fuselage. Maximum cabin pressure differential is 5.5 psi. To ensure clearance for the big propellers long undercarriage legs have been provided with single mainwheels and a double nosewheel.

On the ground and in the air the Jetstream 31 is a strikingly handsome aircraft.

Cabin and flight deck
A very large door at the rear of the cabin lets down from the left side of the fuselage. There are the usual built-in stairs and handrail, but the Jetstream's are more akin to an airliner than a light twin. Directly opposite the entrance door is another which opens onto a toilet and washbasin. The door opens left and locks into a frame in the rear cabin bulkhead. This is a good arrangement because, when in use, the entire vestibule area becomes part of the wash-and-brush-up department.

The Jetstream is offered in three versions: there is the Commuter with eighteen or nineteen passenger seats arranged two on the right and one on the left with an offset aisle, the Executive Shuttle and its twelve posh armchairs positioned six each side of a centre aisle, and the Corporate Aircraft, a ritzy nine- or ten-seat corporate turboprop. All of them enjoy the largest cabin in the small turboprop business. It is 6 ft high, and 6 ft, 1 in wide. No other turboprop in the eighteen- to twenty- seat class can offer such standards of comfort. Furnishings are built to withstand airline service, and passengers who have never flown in anything smaller than a DC9 or a BAC 111 will relax and be at ease in the Jetstream. It has the solid, comforting feel of a big aircraft.

British Aerospace Jetstream 31, the smallest light turboprop with a proper stand-up cabin.

The flight deck is entered through a door in the forward cabin bulkhead. It is large, beautifully planned and completely uncluttered. Ice protection, starting and fuel management are grouped on a roof panel, each pilot has a set of flight instruments with the avionics and weather radar between the two. Engine instruments are arranged in the now almost standard two columns, one for each engine, and these are to the right of the captain's flight panel. Being single-shaft engines the Garretts have only one rpm indicator that reads in percentages of maximum propeller speed. The torque meters are likewise calibrated in percentages. There are the usual ITT gauges, fuel flow meters and combined oil pressure/oil temperature gauges.

The glareshield carries a comprehensive annunciator panel and a wide console extends down from the centre of the panel then back between the pilots. From front to rear it contains the pressurization controls, undercarriage lever, parking brake, air-conditioning and flap position indicator. Immediately behind are two power levers and two propeller levers with pull-up feather toggles which also function as fuel cut-offs in an emergency. It is not possible to incorporate fuel cut-offs with the propeller lever because electric starters on single-shaft motors have a lot of work to do spinning the propeller as well as the engine, and feathered blades would be out of the question.

Behind the engine controls are the three trim wheels (elevators, ailerons and rudder) along with their position indicators.

On a ledge within convenient reach of the pilot's left hand is the power steering knob. The windscreen consists of seven panels and there are too many thick pillars for my liking. On the other side of the penny, there is a lot to be said for optically flat, distortion-free windows that have been stressed to withstand a bird strike.

Electric rocker-type switches are on two conveniently angled panels below the flight instruments. The flight deck has clearly been designed in consultation with pilots who know about flying the commuter routes. Everything falls nicely to hand and aesthetically the 'office' equals the best I have seen. Everything is provided yet it looks incredibly uncomplicated. And the pilots' seats are as good as you can get in a small aircraft.

In the air

Starting sequence is completely automatic once the start switch has been thrown. There is nothing the pilot must do other than monitor the ITT and ensure the temperature stays below limits. At about 15 per cent rpm the ignition does a light-up job on the fuel that starts spraying without need to open the fuel cut-off, then the engine (and propeller) spools up until at 55 per cent the motor is self-sustaining. The ignition switches itself off, the starter

converts to a generator and we are in business.

Nosewheel steering is not one of the aircraft's finer features. It is heavy and not all that accurate. BAe are aware of this and I understand the system is being modified for production aircraft. The Bandeirante system, described in the previous airtest, is very much better in every respect.

Turboprop engines are vulnerable to overtorquing (that is, too much power for the well-being of the propeller transmission), particularly during take-off, and overheating at altitude. Cook an engine and it will be the most expensive fry-up you will ever make. The Garrett engines are fitted with torque and temperature limiters so you can firewall the power levers without having to watch tiny dials at a time when everything is happening. At maximum weight, V_r was a modest enough 105 kt and V_2 108 kt.

I released the brakes, steering with the left hand until 70 kt was indicated, at which point the rudder became effective. At V_r the nose was raised to about 15 degrees and we left the ground, accelerating rapidly to the 130–140 kt initial climbing speed with the VSI showing 2,500 ft/min. The aircraft was near its maximum weight at the time. We were not more than 100 ft above the runway when the BAe test pilot on my right pulled the left power lever back to the stops. Little rudder pressure was needed to check the ensuing yaw. Single-engine climb is in excess of 500 ft/min (double that at average operating weights).

Ideal altitude band for the Jetstream 31 is 18,000–25,000 ft, where cruising speeds vary between 235 and 271 kt according to the power settings used. At 8,000 ft, 97 per cent propeller rpm and 80 per cent torque produced a useful 254-kt TAS. Noise levels are very low for a propeller-driven aircraft and customers in the double-Scotch area are going to enjoy those big oval windows, all that space and the general lack of vibration.

Handling is excellent. Earlier Jetstreams had a single servo tab on one aileron and lateral control was guaranteed to develop the pilot's muscles. The latest model, with its dual servos (one on each aileron) is a great improvement. Stability in pitch and yaw are first class but laterally there is little tendency for the wings to roll level without help from the pilot, which is a pity. It is difficult to assess the stall because there is a stick-pusher that will catapult through the windscreen those who dare resist, but I

OPPOSITE: Commuter version of the BAe Jetstream 31. There is no shortage of cabin space.

BELOW: Passengers are bound to be impressed by the solid feel of the Jetstream, which is built to airline standards of engineering.

recorded stick activation of 95 kt clean and 80 kt with flap 50 degrees.

The correct circuit and approach technique is to set 97 per cent propeller rpm and 30 per cent torque, then reduce speed to maintain height. Flap 10 degrees results in an IAS of 130 kt which is ideal for base leg. When the wheels are lowered the additional drag is countered by turning off the pressurization. That provides just enough power to balance the effect of dangling the wheels in the breeze.

Initial approach is flown at 120 kt, adding flap in stages to arrive ready for landing at 105 kt. You must leave on power until the round-out then the power levers come back to the stops, she sinks to the runway, you lower the nosewheel, apply lift-dump to glue the wheels to the airport then lift the power lever latches and hit reverse thrust along with the brakes, grabbing the nosewheel steering at 70 kt. It sounds like a job for a one-armed paper-hanger but this is a two-crew aircraft and the Jetstream is really quite easy to land.

Capabilities

You can carry eighteen passengers and their baggage over three 155-nm sectors without refuelling. Alternatively, eleven customers can be flown on a 35-nm sector (there is no landing-weight restriction) followed by seven 110-nm sectors representing eight take-offs and eight landings without refuelling. The twelve-seat Executive Shuttle will fly its full cabin on five 125-nm sectors or three 260-nm sectors – all without refuelling. Alternatively, you can fly the full cabin for a nonstop 1,000 nm. Ferry range is almost 1,300 nm. These ranges allow for a 100-nm diversion and a 45-minute hold. Operating weight (equipped empty weight plus crew) leaves a useful load of 5,614 lb (2,547 kg) so it is a flexible aircraft. 4,182 lb (1,897 kg) can be put in the cabin, which is enough for nineteen passengers and a lot of baggage – more, in fact, than the airline allowance. At 57.6 passenger nm/Imp gal, the Jetstream is even more fuel-effective than the Embraer Bandeirante (see previous airtest). This translates into 48 passenger nm/US gal and 12.67/litre.

Field performance, computed on a public transport basis, requires a 4,000-ft runway when operating at maximum weight.

Verdict

Strange as it may seem, although the Jetstream 31 was introduced twenty years after the original Handley Page models first appeared, at time of publication it is the only eighteen- or nineteen-passenger turboprop with a proper stand-up, pressurized cabin. No rival is even on paper. Its chances of success today are probably higher than they were in the 1960s because since that time there has been a remarkable growth in commuter airline operations. Many of these outfits are small and to them a plane stuck on the ground for want of spares is the kiss of death. British Aerospace are marketing the Jetstream 31 like an airliner with twenty-four-hour service and product support

to match.

A few years back a well-known regional airline in the United States called in a firm of business consultants to report on the state of their company. All assets were carefully examined and their Jetstream fleet, some of them fifteen-year-old examples built by Handley Page, was pronounced 'as new'. I regard the Jetstream 31 as an outstanding aircraft in every respect.

Facts and Figures

Dimensions

Wing span	52 ft
Wing area	274 sq ft
Length	47 ft, 2 in
Height	17 ft, 6 in
Cabin length	24 ft, 3 in
Height	6 ft
Width	6 ft, 1 in

Weights & loadings

Max ramp	14,660 lb (6,650 kg)
Max take-off and landing	14,550 lb (6,600 kg)
Max zero fuel	13,228 lb (6,000 kg)
Operating empty (with crew)	9,046 lb (4,103 kg)
Useful load	5,614 lb (2,547 kg)
Max fuel	385 Imp/465 US gal (1,760 litres)
Seating capacity (with crew)	21
Max baggage	900 lb (2,547 kg)
Wing loading	53 lb/sq ft
Power loading	8.08 lb/hp

Performance

High-speed cruise	246–270 kt
Long-range cruise	219–236 kt
Maximum range	1,300 nm
Rate of climb	
Two engines	2,230 ft/min
One engine	520 ft/min
Service ceiling	
Two engines (max operating)	25,000 ft
One engine	12,000 ft
Take-off distance over 50 ft (BCAR)	4,000 ft
Landing distance over 50 ft (BCAR)	4,000 ft

Engines

2 × Garrett TPE 331-10 single-shaft turboprops each developing 900 shp, driving Dowty-Rotol 106-in (4.06-m) diameter, four-blade, advanced propellers.

TBO

6,000 hours or 'on condition'.

Shorts 330

The flying box from Belfast

Background

Even before the Wright brothers had made their first leap off the ground, Eustace and Oswald Short were in the aviation business manufacturing balloons. They had a small workshop at Battersea, London, and in 1905 supplied three reconnaissance balloons for the Indian Army. It was during Wilbur Wright's exhibition flights while in France that Oswald told brother Eustace: 'This is the finish of ballooning.' So they persuaded their oldest brother, Horace, to join in building Wright Flyers under licence. Thus was born the world's first aircraft manufacturing company in November, 1908.

The impressive Short brothers were to be first in the field on many occasions. In 1920 their Silver Streak biplane was the first aircraft with a stressed-skin airframe; the Singapore of 1926 was the world's first large flying boat with a metal hull; the Stirling was in 1939 the first four-engine heavy bomber to enter RAF service; and in 1957 the Shorts SCI was the world's first jet-powered VTOL aircraft. On a less happy note their historic company, which produced some of the finest civil and military flying boats of all time (the C class and the famous wartime Sunderland) was the first to be nationalized, but the less said about that the better. They now operate from a massive factory in Belfast, Northern Ireland.

One of their post-war projects was the Skyvan light freighter. This odd-looking bird, with a cabin cross-section of 6 ft, 6 in (1.83 m) by 6 ft, 6 in, was in contrast to its fat fuselage fitted with a very high-aspect-ratio wing of the type pioneered by Avions Hurel-Dubois in France. It was almost 65 ft in span. Skyvans are still in production and they can lift a 4,600-lb (2,087-kg) payload, climbing 50 ft above the runway after a take-off distance of only 1,100 ft.

In the United States a little piece of highly practical legislation called deregulation was introduced during 1972. In effect this allowed the fast-growing commuter operators to fly bigger aircraft than the previous limit of fifteen to twenty passengers. Strangely the American manufacturers were slow to react and no suitable commuter aircraft were forthcoming. To their credit, Shorts did a market survey and in 1974 came up with the idea of a stretched Skyvan with thirty passenger seats, more power and a minimum of fresh design work. They called it the Shorts 330 and its success has led to a further stretch, the model 360, which is now offered as an alternative to those regional airlines wanting thirty-six seats. This test report deals with the Shorts 330.

Shorts 330. The mainwheels retract into large spats and the nosewheel lifts into a wheel-well under the front fuselage.

Engineering and design features

With a wing span nudging 75 ft and a weight of 22,900 lb (10,387 kg) the Shorts 330 is on the edge of the big aircraft class. The wing outer panels are more or less off the Skyvan. They attach to a large centre-section which carries the engines. Stub wings extend from the bottom of the fuselage and terminate in a large wheel spat at each end. The mainwheels retract into these spats and the nose-wheel disappears upwards into the nose of the aircraft. The outer wing panels are braced by single wide-chord struts which extend upwards from the wheel spats. Each strut comprises three separate tension members so that a fatigue crack in any one will not put the aircraft at risk. Frise ailerons are fitted and the slotted flaps are in six sections. Aspect ratio is a remarkable 12:1. Pneumatic boots provide ice protection for the flying surfaces.

Fuel is carried in two main tanks; one in front of, the other behind, the wing as it passes over the cabin area. Fairings are added to give the fuselage a marked, humpbacked appearance. The top of the fuselage probably contributes a considerable amount of lift. Maximum fuel capacity is 480 Imp/575 US gal (2,176 litres).

The tail surfaces are borrowed from the Skyvan and

ABOVE: Slow-running, five-blade propellers ensure low noise levels.

OPPOSITE: The Shorts 330 has gained a following with commuter airlines in many countries.

consist of a very small tailplane with a large rectangular fin and rudder mounted on each end. There is no fuselage width reduction from behind the flight deck to the tailplane but the under-surface of the rear fuselage sweeps up sharply to meet the top of the aircraft. Fuselage construction is based upon square-section Skyvan frames, giving the same-size 6 ft, 6 in (1.83 m) high, 6 ft, 6 in wide cabin. The fuselage skins are like corrugated paper with a light-alloy fluted inner skin bonded to a smooth outer surface – an unusual method of construction that results in a very rigid structure.

With so many years' experience in building flying boats and float planes anti-corrosive protection is a point of honour at Shorts. Great care is taken to safeguard the 330 at every stage of construction. This is a long-life airframe; over 200,000 flight cycles having been demonstrated under test.

A baggage loading ramp lets down at the rear and more

luggage can be stowed in the nose. Total capacity is 1,000 lb (454 kg).

As already mentioned there is a retractable undercarriage, although the mainwheels require large spats. I suppose they must have somewhere to go but it is a pity they could not hide under the fuselage. Shorts assure me that significant performance improvements result from tucking the wheels into these spats – faster cruising speed, better rate of climb, and a 600-lb (272-kg) increase in payload, for example.

Although this is not a pressurized aircraft Hamilton Standard air-conditioning is fitted. Electric current is provided at 28 volts DC by the starter/generators, and

inverters convert some of this to 115 volts AC for the avionics and instruments.

The engines are PT6A-45R free-turbine units developing 1,198 shp and driving slow-running, five-blade propellers. From some angles the Shorts 330 looks like the shape of things to come as we would have forecast them in the 1930s. Certainly it is different!

Cabin and flight deck

There is a large double door at the front of the cabin and another single one at the rear. They are located on the left-hand side of the aircraft. Ten rows of seats, two on the right and one on the left separated by an offset aisle are serviced by the usual airline-type overhead lockers, their undersides housing recessed passenger amenity panels (reading light, fresh-air vent and hostess/steward call button). At the rear of the cabin is a full-size airline toilet, a coat wardrobe, a small galley for beverages and light meals, and a seat for the cabin attendant. The seats may quickly be removed to convert the 330 into a freighter with a 7,500-lb (3,402-kg) capacity. This is a big, bright and airy cabin for thirty people.

The flight deck is enormous and visibility through the various flat screens is first rate. There is an excellent roof panel where the fuel and electrics are managed. The system is laid out in pictorial form with small magnetic indicators that flick through 90 degrees like railway signals to depict which sections are open and which are closed. Collins flight director systems are standard equipment and the main instrument panel is laid out with the engine dials in the centre between the two flight panels, with the radio installation down below the engine readouts.

The engine controls comprise power levers, propeller levers and fuel condition levers. Nosewheel steering is mounted on the left sidewall. It takes the form of a small lever that moves forward for right turn and back to turn left. They have the same arrangement on the Skyvan, which has the *worst* nosewheel steering I have experienced on *any* aircraft. It works rather better on the 330.

Commuter pilots faced with a day flying short routes punctuated by countless take-offs and landings cannot fail to appreciate this roomy, well-planned flight deck which is something of a home-from-home, with everything falling conveniently to hand within reach of the comfortable crew seats.

In the air

The PT6 engines are started like those of any other turboprop and when time comes to move off, the propellers are unfeathered. The nosewheel steering could do with being a little lighter but works well enough; the lever suspension undercarriage gives a very comfortable ride; taxiing is perfectly straightforward; and the brakes are very good indeed. The flaps are controlled on a gated selector marked 4°, 8°, 15°, and 35°. Flap 4 degrees is used for take-off. At average weights V_1 V_r (decision speed/rotate speed) is 90 kt and V_2 (take-off safety speed) is only 2 kt faster. These are light-twin speeds.

With the propellers set to maximum rpm the power levers were advanced to 3,500 ft:lb torque and the aircraft accelerated briskly past 45 kt, at which point the twin rudders were able to take over from the nosewheel steering. At 90 kt I eased back on the control wheel, left the runway, raised the wheels and went for the 139-kt en-route climb as I lifted the flaps. Rate of climb was around 1,000 ft/min but higher rates can be achieved if you adopt a slightly lower speed.

At 8,000 ft, which would be a typical cruising level for commuter operations in this aircraft, 1,200 propeller rpm and 3,600 ft:lb torque gave an IAS of 170 kt, which trued

out at the manufacturer's claimed 195 kt. The 330 is remarkably quiet internally (which is good for the passengers and crew) and externally. That too is important because it makes the Shorts 300 socially acceptable at those downtown airstrips where the locals have to live under the take-off path of constant traffic.

Stability is so good that many operators do not fit an autopilot since it is considered unnecessary for short-route flying. Handling is ideal for this class of aircraft with adequate aileron response – which is surprising considering the span of the 330 and the fact that simple

Facts and figures

Dimensions

Wing span	74 ft, 8 in
Wing area	453 sq ft
Length	58 ft
Height	16 ft, 3in
Cabin length	31 ft, 1 in
Height	6 ft, 6 in
Width	6 ft, 6 in

Weights & loadings

Max take-off	22,900 lb (10,387 kg)
Max landing	22,600 lb (10,251 kg)
Operating empty (ie, with crew)	15,000 lb (6,804 kg)
Useful load	7,900 lb (3,583 kg)
Seating capacity (with crew)	33
Max baggage	1,000 lb (454 kg)
Max fuel	480 Imp/575 US gal (2,176 litres)
Wing loading	50.55 lb/sq ft
Power loading	9.56 lb/hp

Performance

High-speed cruise, 10,000 ft	190 kt
Long-range cruise, 10,000 ft	157 kt
Maximum range at 190 kt	700 nm
Maximum range at 157 kt	840 nm
Rate of climb	
Two engines	1,180 ft/min
One engine	300 ft/min
Take-off distance over 50 ft	3,900 ft
Landing distance over 50 ft	3,900 ft
Service ceiling	
Two engines	23,000 ft
One engine	11,300 ft

Engines

2 × Pratt & Whitney of Canada Ltd PT6A-45R developing 1,198 shp (flat-rated), driving Hartzell 108-in (2.74-m) diameter, high-activity, five-blade, reverse-pitch propellers.

TBO

3,000 hours.

manual controls are fitted. Likewise the stall is more gentle than many a light plane, 78 kt flaps-up and 66 kt when maximum flaps have been set. Rudder forces needed to check the yaw following an engine failure make modest demands on the pilot and only one-eighth of a turn was needed on the rudder trim to regain balanced flight. I would regard the 300 ft/min single-engine climb (at maximum weight) as acceptable but nothing to write home about.

On the day of my flight the visibility began to deteriorate at Belfast Airport so my first landing in a 330 entailed an ILS approach. An ideal speed for this is 120 kt, going for 85 kt over the threshold. Speed stability is near-perfect and the aircraft is very easy to fly on instruments. The Shorts 330 is no more difficult to land than a light plane, and even without a reverse thrust, landing distance at maximum landing weight is only just over 4,000 ft.

Capabilities

Operating weight (that is, fully equipped with two pilots and a cabin attendant but without usable fuel) is typically 15,000 lb (6,804 kg). Thirty passengers and their baggage would represent a payload of 5,790 lb (2,626 kg), leaving 2,100 lb (953 kg) for fuel, that is, 262 Imp/315 US gal (1,192 litres). You need 775 lb (352 kg) for a 45-minute hold and a 50-nm diversion and with these reserves the full cabin just described could be flown for 270 nm at 190 kt or 300 nm at 157 kt. With full tanks, twenty-one passengers could be flown for more than 700 nm at 190 kt but the Shorts 330 is essentially a short-range commuter plane. With thirty seats occupied it will return almost 50 passenger nm/Imp gal, 41 nm/US gal or 10.87 nm/litre, which is good for the balance sheet.

Verdict

The Shorts 330 is being used in Canada, the United States, Thailand, West Germany, Britain and Australia, to name but a few of its areas of operation. One of the longest routes, Bangkok to Khon Kaen is part of the Thai Airways network, but in the main most of the sectors are under 200 nm, which is typical of commuter services.

The newer Shorts 360 has a tapered rear fuselage, a single swept fin and a stretched cabin, and is consequently a better-looking aircraft. It will be interesting to see if demand for another six seats and improved nautical miles per pound of fuel sets the stage for the demise of the 330. So far there appears to be an interest in both aircraft but that is understandable. As a pair, they are unique in their class.

HOW DO THEY RATE?

The following assessments compare aircraft of similar class. For example, four stars (Above average) means when rated against other designs of the same type.

★★★★★ Exceptional
★★★★ Above average
★★★ Average
★★ Below average
★ Unacceptable

Aircraft type	Appearance	Engineering	Comfort	Noise level	Visibility	Handling	Payload/ Range	Field Performance	Cruising Speed	Economy	Single-engine Climb	Value for Money
Pilatus PC-6 Turbo Porter	★★	★★	★★★	★★★	★★★	★★	★★★★	★★★★★	★★★	★★★★	n/a	★★★
Cessna 425 Corsair/ Conquest I	★★★★	★★★	★★★★	★★★★	★★★★	★★★★	★★★	★★★★	★★★	★★★	★★★	★★★
DHC6 Twin Otter	★★★	★★★★	★★★	★★	★★★★	★★★★	★★★★	★★★★★	★★★	★★★★	★★	★★★★
Piper PA-42 Cheyenne III	★★★	★★★	★★★★	★★★★	★★★★	★★★★	★★★	★★★	★★★★	★★★	★★★★	★★★★
Beechcraft Super King Air B200	★★★	★★★★	★★★★	★★★	★★★★	★★★	★★★★	★★★	★★★★	★★★★	★★★★	★★★
Embraer EMB-110 P2 Bandeirante	★★★	★★★★	★★★★	★★	★★★★	★★★★	★★★★	★★★★	★★★	★★★★	★★★	★★★
British Aerospace Jetstream 31	★★★★	★★★★★	★★★★★	★★★★	★★★★	★★★★	★★★★	★★★	★★★★	★★★★	★★★★	★★★★
Shorts 330	★★★	★★★★	★★★★★	★★★★	★★★★	★★★	★★★	★★★★	★★★	★★★★	★★	★★★★

7. JET AIRCRAFT

The technology

Although the Germans were first by a very short head to get a jet fighter into the air, they paid the price of aiming too high. From the start their engines had axial flow compressors, whereas Frank Whittle in Britain was content to walk first and run later. His early engines used well-understood vacuum-cleaner technology and made do with centrifugal compressors. As a result the Germans never produced a wartime jet engine that would run reliably for more than a few hours. On the other hand, the Gloster Meteor, which appeared soon after the Messerschmitt 262, provided fairly reliable service from the start.

The jet engine is delightfully simple, very much more so than some turboprop units. Air enters the front of the engine, is compressed and then directed through a combustion chamber where burning fuel endows it with high energy. From here it escapes through the turbine section, enabling it to fulfil its main purpose which is to drive the compressor through a shaft running forward to the front of the engine. Thrust is, in fact, the reaction to the mass of high-speed gas being expelled from the tailpipe.

This sort of engine, called a turbojet, at one time powered all jets, civil and military. However, the trouble with turbojets is that they are noisy and thirsty, particularly at low cruising levels.

By the early 1950s the jet engine, which by then had spawned the turbopropeller unit (see previous chapter), took a step in a rather similar direction. A turbine drives the compressor stages as before but now a second turbine section makes use of the exhaust gasses to drive through suitable reduction gears a fan at the front of the engine. The fan is of larger diameter than the gas generator, or core of the engine as it is often named, consequently we now have an interesting situation. The high-speed, hot and very noisy gasses being ejected through the engine tailpipe are surrounded by a tube of cold, relatively slow-moving air. It is rather like striped toothpaste on a grand scale but the arrangement brings with it two benefits of great value: there is a fuel saving of 50 per cent or more and the tube of cold air acts as sound insulation.

Engines with a separate fan section are appropriately known as fanjets and the latest versions, which sport a fan of perhaps twice the diameter of the hot sections, are incredibly quiet and remarkably fuel efficient. Today's aero engines are developing along two parallel lines. There are the new turboprops with multiple-blade propellers that are getting to look more and more like fans; and the fanjets

with their fan units growing bigger so that they look more and more like multiple-blade propellers. Soon the only difference between the two will be whether or not the fan/propeller runs within a shroud. Meanwhile, turboprops and fanjets remain two distinct families of aero engines; but for how long? Who can say?

What's it like to fly a jet?

There is nothing like a jet. No propeller-driven aircraft, piston or turbine, can equal its smooth, effortless flight, the complete lack of vibration or the utter silence on the flight deck, particularly when the engines are rear mounted. Your first take-off in such a jet is a never-to-be forgotten experience. You open the throttles (usually known as thrust levers, because that is precisely what they are), the aircraft accelerates as though being pulled forward by a giant invisible magnet, with nothing to break the silence other than the 'thump–thump, thump–thump' of the wheels as they race ever faster across the runway paving.

As you would expect, jet flying demands special techniques, certainly when the aircraft has swept wings, and this applies particularly during the approach and landing. However, it is not the purpose of this book to act as a flying training manual and I shall confine myself to explaining some of the features most likely to interest the prospective user.

What does the fanjet have to offer?

For journeys of up to, say, 750 nm a good turboprop is often adequate, but although some of the larger ones have a range potential of 1,500 nm or more I would not recommend using them for such trips on a regular basis. Six hours spent in even the finest of the general aviation turboprops is not conducive to obtaining the best from business executives who may be required to enter complex negotiations on arrival at their destination.

Other than the smallest and lowest-powered jet on the market, all others offer a considerable jump in cruising performance, but once you enter the realm of company jet travel you meet up with the law of diminishing returns. By that I mean time saved versus speed increases. This is important because some business jets derive, say, a 40–50 kt speed advantage over others by providing a restricted cabin in which you need to be built like one of the seven dwarfs when you go to inspect the plumbing.

Now 40 or 50 knots sounds quite an impressive speed

increase, and so it is when related to a light single or small twin. But look at these figures. They show how long it takes to fly 500 nm at block speeds that increase in steps of 50 kt.

Block speed (kt)	Journey time over 500 nm	Time saved (min)
200	2 hr, 30 min	–
250	2 hr	30
300	1 hr, 40 min	20
350	1 hr, 26 min	14
400	1 hr, 15 min	11
450	1 hr, 7 min	8
500	1 hr	7
1,000	30 min	30

These figures clearly illustrate that once you start exceeding 400 kt, time saved by flying 50 kt faster on a 500-nm journey becomes of little significance, unless you buy a supersonic bizjet and fly at 1,000 kt – but privately owned supersonic jets are a bit thin on the ground. The message I am trying to convey is that if I were a compny chairman faced with making a choice between a 400-kt jet with a big, comfortable cabin and a 450-kt tube of a bizjet. I would go for the slower, more civilized one every time. The eight minutes saved is not worth all those crumpled suits and cramped surroundings.

About the hardware

Jets are happiest cruising rather higher than turboprops, 30–41,000 ft being typical. Consequently they are subjected to larger temperature variations, and while a cabin pressure differential of 5.5 psi or less would be representative of turboprops, 8.5 psi is needed to maintain an 8,000-ft cabin while cruising at 41,000 ft.

Jets burn a lot of fuel because of the power required to fly so fast. Therefore what goes in the tanks, later to disappear out of the jet pipes, represents a high proportion of the total weight. A British Aerospace 125/700, for example, carries up to 9,450 lb (4,287 kg) of fuel (see final airtest in this chapter). And that amounts to more than 38 per cent of the maximum take-off weight of the aircraft. Consequently there is often a considerable landing weight restriction, particularly with the larger jets.

Because of the speeds involved and the widely varying flight conditions, even the smallest jets must be equipped with airline-standard avionics and first-class ice protection. And to meet the very high airworthiness requirements rightly demanded by all countries, it follows that engineering standards must be higher than those of most aircraft of lower performance. All this adds up to an aircraft that is perhaps large for its passenger capacity, relatively complex in its systems and more expensive than other categories, even turboprops.

What should be expected of a good jet?

- It must offer standards of comfort that will justify using it to transport key personnel who might otherwise have gone by scheduled airline.
- Noise levels must be lower than the best turboprops.
- Cruising speeds must be significantly higher than those of the best turboprops to justify the much greater cost of purchase and operation.
- Field performance must allow it to visit modest airfields. One well-known bizjet had airfield demands that precluded its use on many occasions. The main justification for running a company jet is that it will take key personnel to airfields that cannot be served by the airlines.
- It must be quiet enough to be accepted at noise-sensitive airfields, many of them surrounded by the anti-aviation lobby (which has been known to make more noise than the aircraft about which it so frequently complains).
- It must be reliable and hard-wearing because often an aircraft may be shared by two or more companies as a means of deriving full economic value from the high productivity which is a feature of jets.

In varying degrees most of the business jets now on offer fulfil the majority (but not all) of these requirements. It is a question of how well they do it that counts.

Are jets economic?

I am often asked how much a company jet costs to buy and operate. When you mention the purchase price and the costs per flying hour the usual reaction is one of utter amazement that any company, however large, should be so foolhardy as to let themselves in for such mind-blowing expense. The trouble is that most people never stop to consider the full implications. In the first place only a company with an annual travelling commitment of at least 200,000 nm should consider buying a jet for their sole use (say, 500 hours flying per annum). Provided a reasonable number of people occupy the aircraft on such business trips, the cost per passenger mile will often work out cheaper than that of a chauffeur-driven company car. But it does not end there. Think of the money saved in hours wasted by expensive executives hanging around the airports of the world waiting for connections, those mammoth hotel bills (when a company jet could have been there and back in a day), the ability to visit several widely spaced destinations in one day, and the prestige effect on potential customers of arriving in the firm's jet. Whatever the arguments relating to running costs, the final three airtests in this book represent some of the finest flying hardware available to the jetset businesses of today.

Cessna 550 Citation II

The owner-driver jet

Background

The early business jets, introduced in the mid-1950s, were large, fast, heavy and, in some cases, hot rods. One even had four rear-mounted engines. Morane Saulnier, the French manufacturers, went the other way and produced a four-seat minijet called the Paris which was built in some numbers. The late Shah of Iran had one and used to beat up the local countryside, until his ministers prevailed on him not to risk his neck.

Piper has not entered the jet market and neither has Beech (except for a short-lived arrangement with Hawker Siddeley when they tried to market the British-built HS125 bizjet). Then in the mid-1970s Cessna came out with a little business jet called the Citation I. It flies on a pair of tiny fanjets and will transport six passengers at 350 kt. Now that may not sound particularly fast in jet terms but it is 70–100 kt more rapid than most turboprops. To give you some idea of what can be done with a 350-kt jet, when I flew the Citation I for some magazines a few years back eight of us (six in the passenger cabin and two in the driving seats) left Gatwick Airport, near London, aimed for the south of France, landed at Cannes, which is 570 nm from home, had a slap-up lunch, flew back to London and got home in time for tea. The Citation I is a great little aircraft but it was not long before Cessna did a stretch, added a little more power, put in two more seats and came up with the Citation II. The first models appeared in 1978. In my opinion it is even better than little brother, although both models are in production and the Citation I continues to have a good following.

The Citation I is light enough for single-crew operation; furthermore, you can certify the Citation II at 12,500 lb (5,670 kg) instead of its potential 13,300 lb (6,033 kg) and enjoy the same privileges. The aircraft described in this airtest report is a Citation II operated at the higher weight, when it must be flown by two pilots.

Engineering and design features

The high-aspect-ratio wing (8.28) has no sweepback and from some angles the Citation II has the look of a sailplane. Apart from the fact that it is not fast enough to warrant a swept wing there are advantages to having a straight one: handling is similar to a piston twin or light turboprop, field performance is modest, and a less obvious benefit is the ability to cruise high on low power. The Citation II can fly at 43,000 ft when required, which is above most or, in some parts of the world, all of the weather.

Slotted flaps have a maximum deflection of 40 degrees and like all jets, which cannot rely on the drag of throttled-back engines and windmilling propellers when time comes to slow down, speed brakes are fitted. These hydraulically operated surfaces take the form of perforated plates which stand up on the wings inboard of the ailerons. They may be operated at any speed. Integral tanks in the wings have a fuel capacity of 620 Imp/745 US gal (2,820 litres).

The fin has a dorsal area that extends forward to the rear of the cabin and mounted on it, a foot or so above the fuselage, is the tailplane, which has pronounced dihedral to ensure that it is kept out of the hot jets from the engines. All flying controls are manually operated via stainless-steel cables.

The fuselage is virtually a circular-section tube. Six large windows are provided on each side of the cabin, there are two enormous ones on either side of the flight deck, and the pilots have a deep two-piece windscreen that allows horizontal visibility through 340 degrees and a total of 52 degrees above and below the forward line of sight. An entrance door behind the flight deck on the left-hand side opens by swinging forward and a two-step ladder is then lowered to assist entry.

Power is supplied by two Pratt & Whitney of Canada Ltd JT15 D-4 fanjet engines, each of 2,500 lb static thrust. Being two shaft units (one for the gas generator, the other running from the power turbine to the reduction gears that drive the fan), two rpm indicators are required. Like all of the smaller jets with rear-mounted engines, those on the Citation II have their air intakes positioned above the wing to reduce the risk of ingesting stones or other debris while manoeuvring on the ground.

The cabin is pressurized to a maximum of 8.7 psi which ensures a sea-level environment when the aircraft is at almost 23,000 ft and an 8,000-ft cabin while flying at 43,000 ft. Pressurization and air-conditioning are supplied by the engines.

Comprehensive anti-icing and de-icing equipment is fitted as standard, consisting of electric heaters on the wing leading edges ahead of the engines, along with rubber boots outboard of these areas and on the fin and tailplane leading edges. Bleed air from the compressors is used to protect the engine inlets as well as the pilots' windshields, which also have standby alcohol spray.

Included in the basic price of the aircraft is a Sperry 'two-cue' flight-director system, with a light plane-type flight panel for the copilot. The radio installation comprises two VHF communication sets, two VHF nav receivers, ADF, transponder, weather radar and an altitude alerter that goes 'B-O-I-N-G' if you depart from your assigned flight level.

The Citation II is not the prettiest of business jets, but it will grow on you provided you accept that it is basically a tube with a long, thin wing.

Cabin and flight deck

Being a circular-section fuselage, the 59-in (150-cm) maximum width coincides happily with your shoulders when seated. An aisle in the centre of the cabin drops 5 in (13 cm) below floor level and provides 57 in (1.45 m) headroom. From behind the pilots to the baggage/toilet bay is 11 ft, 4 in (3.45 m). Various layouts are available but the example flown had the so-called 'double-club' seating: two facing backwards behind the pilots, another two opposite, looking forward. Behind them are two more facing rear, with another pair opposite. The seats are comfortable but not outstanding. The entrance opens between the front two pairs and a generous central aisle allows easy movement from one end of the cabin to the other. Behind the rear seats in the baggage area is an electrically flushing toilet but no washbasin, so you would need impregnated paper towels. There is room for a good refreshment centre and an icebox. There are fold-out tables and all the usual reading lights, adjustable fresh-air vents and airline-type emergency oxygen masks that emerge from the ceiling in the event of depressurization. A small wardrobe completes this picture of domestic bliss in the air.

Cessna have a talent for good flight decks in their light twins and turboprops. The one in the Citation II is very similar to that of the Citation I, and I think Cessna have done a first class job in both. Circuit breakers for the avionics and electric instruments are on a right sidewall panel, those for most of the electric services being on the left. Electric and fuel management are to the left of the captain's flight panel, there is a control lock under the left-hand yoke and an emergency brake lever nearby. The normal brakes are powered and have anti-skid facility.

The centre of the panel contains a row of vertical-reading engine instruments. From left to right they indicate fan rpm (known as N_1 turbine) interstage turbine temperature (ITT) and fuel flow in pounds per hour. Below them are rpm indicators for the gas generators (N_2 turbine), which take the form of digital readouts without vertical scales. Next to the fuel flows are vertical-reading fuel contents gauges calibrated in pounds, followed by oil temperature and oil pressure gauges. While on the face of it the idea of looking at a row of thermometer-like readouts might sound attractive I am not entirely sold on these vertical instruments. The fact that electronic digital readouts are incorporated to supplement the vertical scales confirms my impression that they are not all that easy to set accurately.

Below the engine instruments is a comprehensive annunciator with a multi-position test switch. You twist this before starting, to be rewarded by a series of multi-coloured lights, some of them accompanied by bells, horns and buzzers – all very entertaining. The centre panel also houses the radio and the weather radar. There is an excellent altitude alerter, which provides a datum for the autopilot when it is operated in the 'altitude lock' mode.

Between the pilots is a central pedestal which houses the simple cabin-pressure controls, the two thrust levers (throttles), the flap lever and a switch for the airbrakes. Elevator, aileron and rudder trim wheels are nearby and so is the autopilot control panel. The example flown also had Omega long-range area navigation. Excellent, all-adjusting crew seats make up the crowning glory of this splendid flight deck.

Cessna Citation II, stretched version of the lower-powered Citation I. This aircraft represents a link between the light turboprops and faster, more complex business jets.

The high-aspect-ratio wing has no sweepback and provides good high-level cruising capabilities for the low power. It also ensures modest field demands.

In the air

To wind up the elastic you turn on the ignition, press the start button, wait for 10 per cent fan rpm to be indicated, then move the thrust lever forward and out of its fuel cut-off position. Fuel then enters the engine, the ITT shows a rise confirming light-up and the turbines rapidly speed up to become self-sustaining. We had 3,400 lb (1,542 kg) of fuel in the tanks, four people seated and enough payload going spare for another six adults and 680 lbs (309 kg) of baggage.

Although the aircraft is heavier than a Twin Otter or a Bandeirante (see Chapter 6), Cessna have managed to provide simple nosewheel steering that contrives to be lighter than some of the piston-twins without need of power assistance. The brakes are excellent, so is visibility, and provided you remember the 51-ft wing, taxiing is no more difficult than in a club trainer. For the take-off at medium weights, speeds to bear in mind are:

V_1 (decision speed)	96 kt
V_r (rotate speed)	100 kt
V_2 (safety speed)	107 kt
V_{yse} (blue-line speed)	142 kt

The cabin is adequate for short journeys and offers similar internal measurements to the smaller Lear Jets.

I lined up, moved the thrust levers fully forward and the little jet surged ahead, reaching 100 kt before you could shout 'Help!' At 220 kt the VSI said we were going up at 3,000 ft/min (in near silence) as I hastily removed myself from the London TMA and headed for Bedford Radar. In a very short time we were at 31,000 ft, where the high-speed cruise gave an IAS of 230 kt. Outside the temperature was minus 38°C. Inside I had my jacket off and we had to turn down the heat. Airspeed trued out at 380 kt and the fuel burn was 600 lb (272 kg)/hr for each engine which, in total, is 150 Imp/180 US gal (681 litres) per hour.

The ailerons are slightly heavier than those of a Citation I and the rate of roll is slower, but all controls, trimmers included, function well. Flaps-up the stall came at 95 kt followed by a gentle roll to the right. Maximum flap reduced stalling speed to 80 kt. In the unlikely event of lost cabin pressure an emergency descent is made by extending the airbrakes then lowering the nose to maintain 240 kt, while the VSI shows 6,000 ft/min DOWN. Asymmetrics present few if any problems, with a single-engine climb rate of over 900 ft/min.

Initial approach at 120 kt, followed by 100 kt over the threshold are similar speeds to those you would fly in some piston-engine twins. The Citation handles beautifully during the approach, there is a small change of attitude during the round-out, and the landing could not be simpler.

Capabilities
Standard zero fuel weight restricts the cabin to a maximum payload of 2,000 lb (907 kg) – excluding crew – but as an option the Citation II can be ordered with a 1,500-lb (680-kg) increase in cabin load.

Total operating weight, which includes two pilots, averages 7,500 lb (3,402 kg), according to equipment and cabin furnishings. With full fuel the payload is 991 lb (450 kg), which is enough for five passengers and the usual airline baggage allowance. At 33,000 ft you could cover a with-reserve 1,800 nm at around 375–380 kt. With eight passengers and their baggage this reduces to a still-useful 1,550 nm. With such a load the Citation II will return almost 21 seat nm/Imp gal, 17.36 per US gal or 4.58 seat nm/litre. A Piper Cheyenne III turboprop (see Chapter 6) with seven passengers provides 10 per cent more passenger miles per unit of fuel but at 100 kt lower cruising speed. It is the old story – you get what you pay for.

Verdict
The Citation II has a better field performance than some of the turboprops and its initial purchase price is no more than the biggest King Air (see Chapter 6). It is probably easier to fly than some of the aircraft already mentioned in this book and it is very much quieter during all phases of flight.

In terms of quality it meets (but does not surpass) the requirements laid down for jet aircraft at the beginning of this chapter. One hears complaints of interior trim becoming shabby rather quickly and possibly Cessna should improve this, even if they need to charge a little more to do so. That apart, the Citation I and II are outstanding value and ideal for companies wishing to graduate from a turboprop to their first jet.

Facts and figures

Dimensions

Wing span	51 ft, 8 in
Wing area	323 sq ft
Length	47 ft, 2 in
Height	14 ft, 9 in
Cabin height	57 in (145 cm)
Width	59 in (150 cm)
Length (ex flight deck)	16 ft, 2 in

Weights & loadings

Max ramp	13,500 lb (6,124 kg)
Max take-off	13,300 lb (6,033 kg)
Max landing	12,700 lb (5,761 kg)
Max zero fuel Standard	9,500 lb (4,309 kg)
Optional	11,000 lb (4,990 kg)
Operating weight (including two pilots)	7,500 lb (3,402 kg)
Useful load	6,000 lb (2,722 kg)
Seating capacity (with crew)	10/12
Max baggage	1,000 lb (454 kg)
Max fuel	620 Imp/745 US gal (2,820 litres)
Wing loading	41.17 lb/sq ft

Performance

Typical cruising speeds (according to wt & altitude)	350–385 kt
Max range (high-speed cruise)	1,800 nm
Rate of climb Two engines	3,370 ft/min
One engine	1,055 ft/min
Take-off distance over 50 ft	2,990 ft
Landing distance over 50 ft	2,270 ft
Service ceiling Two engines	43,000 ft
One engine	25,200 ft

Engines
2 × Pratt & Whitney of Canada Ltd JT15 D-4 fanjets giving 2,500 lb static thrust.

TBO
3,000 hours.

Dassault-Breguet Mystère Falcon 10

The executive's jet fighter

Background

In 1967 the two French private companies of Marcel Dassault and old-established Breguet Aviation merged to become Avions Marcel Dassault-Breguet Aviation. Compared with such giants as Boeing in the United States, British Aerospace and indeed Aerospatiale of France, Dassault-Breguet with its 15,000 staff is a relatively small outfit. However, more than a quarter of these are employed on design, research and development work and most of the production is contracted out. Dassault-Breguet are renowned for their formidable Mirage fighters. They also partner Dornier of Germany in producing the Alpha-Jet military trainer. Then there is their Super Etendard ship-borne attack aircraft which hit the headlines during Britain's successful campaign to repossess the Falkland Isles from an Argentinian task force in 1982. Dassault-Breguet also co-operated with British Aerospace in building the Jaguar fighter.

Things are not all military at Dassault-Breguet. In 1962, before the already mentioned merger, Dassault began studies on the Mystère Falcon 20 – a business jet slightly larger than the de Havilland 125 which came out some eighteen months earlier in Britain. The Falcon 20 sold well but marketing experience with that handsome aircraft revealed a demand for a smaller version for up to seven passengers, offering high speed and greater economy. The new design was called the Mystère Falcon 10. The prototype flew on 1 December, 1970 and deliveries began late in 1973. Apart from being perhaps the most beautiful business jet flying, it is in many respects the most advanced. It is also the fastest.

Engineering and design features

The fail-safe airframe is constructed in zinc-free aluminium alloy to the highest standards of engineering. There are two spars in the wings, and electrically operated, double-slotted Fowler flaps allied to hydraulic-powered leading-edge slats, the outer sections being controlled by angle-of-attack vanes that automatically move them open when the aircraft is flying at low speed. Ahead of the flaps are the hydraulic airbrakes which may be extended at any speed – and the Falcon 10 has been dived to Mach .98 in the course of which it reached indicated airspeeds of 450 kt. The ailerons, like all flying controls on this aircraft, are powered by duplex hydraulic actuators. One half of the actuators is supplied by the left engine pump, the other half by the right. If an engine fails, the powered controls continue to function at half rate. There is also a standby electro-hydraulic pump for dealing with the unthinkable –

Dassault-Breguet Mystère Falcon 10, smaller than a BAe 125, larger than the Lears and faster than most other jets.

The flight deck is crammed with equipment but all controls fall nicely to hand. The rather complex pressurization management is to the left of the copilot's control yoke.

a double engine failure (when the Falcon 10 would become a very expensive glider). Even if all hydraulic pumps fail it is not the end of the world, because the controls can be operated manually. I am told that at reduced airspeeds they are heavy but manageable. The hydraulics operate the undercarriage, airbrakes, slats, anti-skid wheel brakes, yaw damper and nosewheel steering. Only the flaps are electric. There are no trim tabs, variable feel being provided by springs that are altered in tension through electric trim switches provided for the pilots.

Up to 730 Imp/880 US gal (3,331 litres) of fuel are carried in two integral wing tanks and two feeder tanks situated in the rear fuselage near the engines. You can fill the aircraft through a single pressure point or, when pressure refuelling is not available, two over-wing points. To reduce spanwise airflow at high angles of attack, two wing fences are fitted approximately one-third span from the fuselage.

A very large, highly swept fin carries the all-flying tail-plane (stabilator for the benefit of readers in the United States). It is positioned about 2½ ft above the fuselage to keep it well clear of the jet efflux. The beautifully pro-portioned fuselage has three large oval windows on each side of the cabin, two side windows on the flight deck, and a simple, two-piece flat windscreen. The entrance door is ahead of the wing leading edge on the left-hand side.

The top half hinges up, while the lower portion drops down to provide three steps up to the cabin floor, which in this sleek little private jet fighter is in any case close to the ground.

There are double mainwheels and a single nosewheel, with doors mechanically linked to the undercarriage legs. Anti-icing for the engine intakes and wing leading edges is provided by hot air from both low- and high-pressure turbine stages. Windshield and pressure/static ports are electrically protected; there is no anti-icing for the fin or tailplane because it has proved unnecessary. Pressurization is ducted at a maximum of 8.8 psi which ensures an 8,000-ft cabin while flying at over 45,000 ft. Full air-conditioning may be maintained on the ground by running one engine at idle power. Some airports, though, do not allow extended engine running while waiting for the passengers to arrive.

Electric power is provided by the engine 9-kw starter/generators. All avionics are stacked within the nose, the top half hinging up and revealing their mass of complex technology whenever technicians are called upon to fix a malfunction.

Two Garrett TFE 731-2-IC fanjets, each giving 3,230 lb

static thrust, are installed within rear-mounted pods with their air intakes shielded by the wings from debris while on the ground. Being a relatively small business jet the engines seem large in comparison, but by any standards the Falcon 10 is one of the best-looking aircraft of any category.

Cabin and flight deck

The airstairs lead to a small vestibule area. Interior arrangements are flexible, but typically there would be a coat wardrobe opposite the entrance and a pull-out toilet. There is room for a small refreshment centre. The interior is 58 in (141 cm) wide, 59 in (150 cm) high and 16 ft, 5 in (5 m) long, excluding the flight deck, so if it cannot match the Falcon 20 or a BAe 125/700 (described in the next airtest), the little Falcon 10 nevertheless provides rather more room than a Citation (see previous test report) or most of the Lear Jets. The very comfortable passenger area could have four seats arranged club fashion with tables between, or four in two facing pairs with a two- or three-seat full-width couch behind. At the rear of the cabin is a 264 cu ft (7.47 cu m) baggage hold. Standards of trim and design are better than anything you will see in the first-class section of an airliner.

To the left of the entrance is a pair of double doors leading to the flight deck. Visibility through its flat, distortion-free windows and screens is excellent. A large roof panel contains switches for all the electric management, anti-icing, wipers, engine starting, cabin and flight-deck lights and the various exterior lights (landing, taxiing, anti-collision strobes, and so on). The aircraft flown had magnificent Collins 5-in (13-cm) attitude directors of the 'V' bar type, with their matching HSIs. Each pilot has similar flight panels with an annunciator display above them. In the central area are the avionics and weather radar. Directly above is a large failure-warning panel with lights capable of conveying up to twenty-three pieces of bad news. The usual engine instruments are lined up in two columns (left engine/right engine) just to the right of centre. From top to bottom they read: fuel flow, interstage turbine temperature, N_1 (fan) rpm, N_2 (gas generator) rpm and combined dials giving oil pressure and temperature. Two fuel gauges above the failure-warning panel read total contents.

To the right of the engine instruments is a delightful little panel that has the red and green undercarriage lights alongside a pictorial airfoil with slats and flaps that move against graduated scales to denote settings. The undercarriage selector is directly below. Pressurization and cabin temperature are managed on a series of instruments and controls grouped to the right of the avionics.

A wide console extends back between the pilots. There is a position indicator for the three trimmers, a parking brake lever, an airbrake lever, the flap-slat lever and a small panel for the trim switches. The two thrust levers have safety latches that permit them to be brought back behind the idle stops and into the fuel cut-off position. The engines are started and shut down with the levers in CUT-OFF. At the back of the console are the autopilot controls and mode selectors, emergency slat extension and a series of test buttons for the warning lights. Against each wall is a ledge carrying emergency oxygen masks for the crew. The left one also has the captain's nosewheel steering knob.

This is a very fine flight deck and it has been possible to keep it uncluttered by removing most of the electrics to the roof. It goes without saying that superbly comfortable seats are provided for the lucky crew whose task it is to fly this exciting little jet.

In the air

Having turned on the fuel boosters you press the start button, wait for 10 per cent N_2 (compressor) rpm to indicate, then move the thrust lever out of the CUT-OFF position to allow fuel into the spray system. A rise in ITT confirms that the fire has lighted, the rpm build up rapidly and soon you can throw a switch to convert the starter into a generator. Power from that can now be used to wind up the other engine without having to wait for the battery while it regains its breath.

You have to press down the nosewheel steering knob before it can be moved left or right; the system is both light to the touch and accurate. On the occasion of my flight we had a take-off weight of about 16,500 lb (7,484 kg), representing average operating conditions. Flap 15 degrees, which automatically opens the slats, is used for take-offs, and I was given a V_1/V_r of 112 kt and a V_2 of 118 kt. I lined up, opened the thrust levers, steered on the nosewheel knob until 70 kt, when the rudder became effective, then took charge of the control yoke with my left hand. At 112 kt I raised the nose to 15 degrees on the ADI, we left the ground and accelerated in near silence to the 300-kt climbing speed. Initial climb was about 6,000 ft/min – a little faster than in your Piper Cherokee. At Flight Level 240 (24,000 ft on a standard day) we set up the recommended high-speed cruise, which entails using 94–96 per cent power and that gave us an IAS of 340 kt which trued out at 485 kt. By any standards that is a rapid gallop – no less than 558 old-world mph (900 km/h). And it was all happening in near silence. It is hard to believe that you are doing little more than opening and closing hydraulic valves when you move the controls; the aircraft seems live and responsive and the artificial feel has been set to give perfectly harmonized ailerons, elevators and rudder. So strong is the airframe that no speed restriction is imposed for flights through turbulence. I was encouraged by the Dassault demonstration pilot to pull 60 degree turns, rolling left and right. It handled like a jet fighter. I tried a dive down to 12,000 ft, hitting Mach .89 without any suggestion of compressibility even though TAS was at times around 630 kt, with the rate of descent nudging 20,000 ft/min. The airbrakes work wonders, even at these speeds. Clean stall came at 112 kt and slats plus maximum flap (52 degrees) reduced that to only 80 kt. There was pronounced pre-stall buffet, no wing drop and an instant recovery as the wheel was moved forward. No stick pusher/shaker is fitted since it is considered unnecessary.

Loss of one engine is almost a non-event in the Falcon 10. Single-engine cruise is a hustling 300 kt and one-engine climb is an equally no-loitering 1,500 ft/min.

Handling on the approach is rock-steady with instant response to the flying and engine controls. Arrival over the fence was flown at 108 kt. The round-out was followed by pulling the thrust levers to the idle stops, and after a brief float the mainwheels touched, the nosewheel was lowered and I applied full brake. The runway was wet but the anti-skid brakes prevented any suggestion of wheel locking. The flight had been an exhilarating experience.

On the ground a Falcon 10 seems to be moving at speed. In the air it certainly does.

Capabilities

Maximum take-off weight (there is no ramp weight quoted) is 18,740 lb (8,500 kg), and a well-equipped example would have a useful load of 7,980 lb (3,620 kg), of which 2,140 lb (970 kg) may be put in the cabin which is enough for two crew, seven passengers and a lot of baggage. That would leave 5,840 lb (2,650 kg) for fuel, which is just about maximum. So it is virtually a full-cabin/full-tanks aircraft. Range with full IFR reserves is 1,200 nm at over 480 kt when flying at 31,000 ft, 1,650 nm at 430 kt (41,000 ft), and 1,800 nm when only VFR reserves are required. Typical take-off distance at high weights would be 4,500 ft and the landing distance is 3,750 ft. Such field performance, which reflects the efficiency of the Falcon 10's high-lift devices, enables this very fast business jet to use relatively small airfields, where its low external

noise level makes it socially acceptable. But speeds of 480 kt and more are bought at a cost: with seven passengers on board we are talking of only 17 seat nm/Imp gal, 14.5 seat nm/US gal and 3.82 seat nm/litre.

Verdict

If you want an aircraft with a stand-up cabin, the Falcon 10 is not for you. Neither does it provide the amenities of a washbasin and running water. However, for the short time you are in the aircraft (a 1,000-nm trip would require 2 hr, 10 min from take-off to landing) these luxuries are not so important. As a very fast and, for its speed, economical way to move four to seven executives over distances of up to 1,200 nm the Falcon 10 is probably unbeatable. Of course, it is a relatively complex little aircraft with most of the facilities and high-lift devices of a big passenger jet,

consequently it is very much more expensive that a Citation II or the smaller Lear Jets but rather cheaper than the Israel Aircraft Industries Westwind or the Lear Jet 55.

Dassault/Breguet claim a 1,900-nm VFR range with four passengers for the Falcon 10. It is at its best carrying such a load, but for practical purposes this magnificent jet should be regarded as a fast, 1,500-nm transport.

Facts and figures

Dimensions

Wing span	42 ft, 11 in
Wing area	259 sq ft
Length	45 ft, 5 in
Height	15 ft, 1 in
Cabin height	59 in (150 cm)
Width	58 in (147 cm)
Length (ex flight deck)	16 ft, 4 in

Weights & loadings

Max take-off	18,740 lb (8,500 kg)
Max landing	17,640 lb (8,000 kg)
Max zero fuel	12,900 lb (5,850 kg)
Operating weight (including two pilots)	11,100 lb (5,035 kg)
Useful load	7,640 lb (3,465 kg)
Seating capacity (with crew)	6/9
Max baggage	550 lb (250 kg)
Max fuel	730 Imp/880 US gal (3,331 litres)
Wing loading	72.36 lb/sq ft

Performance

Fast cruise (Mach .83 at 31,000 ft)	482 kt
Long-range cruise (Mach .75 at 41,000 ft)	430 kt
Range at 482 kt (31,000 ft) IFR	1,200 nm
Range at 430 kt (41,000 ft) VFR	1,800 nm
Rate of climb Two engines	4,500 ft/min
One engine	1,500 ft/min
Take-off distance over 50 ft	4,500 ft
Landing distance over 50 ft	3,750 ft
Service ceiling Two engines (max operating)	45,000 ft
One engine	27,400 ft

Engines
2 × Garrett TFE 731–2–IC fanjets giving 3,230 lb static thrust.

TBO
Not limited. Engines maintained on a contractual 'power-by-the-hour' basis.

British Aerospace 125/700

The ultimate mini-airliner

Background

The late Edward Henry Hillman, a rough diamond of a haulage contractor, founded Hillman Airways in England during the early 1930s. One day he burst in on Geoffrey de Havilland with a crude sketch of a twin-engine biplane based on the great manufacturer's Fox Moth and told him to build it (old man Hillman never asked anyone, they were always told). The Hillman brain-child emerged as the DH84 Dragon, a remarkable small 'airliner' that would fly ten passengers on a pair of 130-hp Gipsy Major engines. Even the 'big' airliners only carried thirty-six people in those days. De Havilland later built an even better development, the DH89 Rapide of which more than 500 were sold; then after the war came the Dove, one of the finest light transport planes ever built. Almost thirty years after the Hillman-inspired Dragon, de Havilland, yet to be submerged in the merger passion that gripped post-war British governments, announced plans for a Jet Dragon. It was going to offer high-speed, over-the-weather transport for six passengers. So far, so good. But where de Havilland got it wrong was in regarding their jet as a Dove replacement, because there was room for both aircraft. Think what a modernized Dove might be today with turboprop engines – but that is another story.

For some reason the name Jet Dragon, which so aptly connected the famous old aircraft with its modern, fire-belching great-grandson, was dropped in favour of a rather impersonal DH125. Its first flight was in 1962. Series 1 aircraft were soon followed by series 3, then in 1968 came the model 400 which featured built-in airstairs, an increase in power which allowed the maximum take-off weight to rise from 22,800 lb (10,342 kg) to 23,300 lb (10,569 kg), and the replacement by suppressed aerials of all the wires and attachments that looked like garden rakes.

The most radical improvements came in 1972. By that time the company had been brought into the Hawker Siddeley empire and the new version of de Havilland's business jet was known as the HS125/600. It had a 2-ft fuselage extension which did a lot to improve the breed's rather chunky appearance, an extra pair of windows, two more seats allowing for up to fourteen passengers in the high-density layout, a maximum weight of 25,000 lb (11,340 kg) and higher-powered engines.

The various models so far described were all powered by Rolls-Royce Viper engines. These straight turbojets had originally been designed by Bristol Aero Engines for use in a high-speed, radio-controlled target drone. Then someone suggested that if they beefed up this bit here and that bit there it would make a cracking engine for civil and military use. Of course, the British government, not content with merging all the airframe manufacturers, did a similar exercise on the engine firms. As a result the Viper became a Rolls-Royce engine. It has powered many light fighters, jet trainers and all the DH and HS125s up to the 600 series. With it, the manufacturers pioneered their famous 'power-by-the-hour' scheme, whereby the owner pays a fixed sum based on flying time, in return for which he gets well-maintained engines. The Viper is an excellent and reliable engine but it suffers from the same deficiencies as all the early generation straight turbojets – excessive noise and very high fuel consumption.

A combination of social pressures and rocketing fuel prices encouraged the development of fanjets (briefly described in the introduction to this chapter). So in 1977 British Aerospace – the company that had devoured Hawker Siddeley's aviation interests in yet another State-instigated merger – announced the BAe 125/700. It was, more or less rivet for rivet, a 125/600 airframe fitted with Garrett TFE 731 fanjets of 3,700 lb static thrust in place of 3,750-lb Rolls-Royce Vipers. The effect on the aircraft of fitting modern, fuel-thrifty engines can truly be described as dramatic. Noise level is now so low that the aircraft has become known as the 'Whispering 125'; range on the same tankage as the 600 series has increased from 1,600 nm to almost 2,700 nm; and perhaps the most vivid illustration of these engines' remarkable economy is that a 700 series cruising at 400 kt burns no more fuel than the earlier models use while taxiing.

Engineering and design features

The BAe 125 airframe is built to the same engineering standards as a large passenger jet; consequently a number of the early series 1 examples are still flying after twenty years. For example, by the time the 700 series was introduced in 1977 the fuselage had been tested through 135,000 flight cylces, the wing-fuselage attachment through 200,000 cylces, and the undercarriage 150,000 flight cycles. The 700 series has a design life which in practical terms amounts to a one-hour flight every day for fifty-five years.

The one-piece wing has three spars, front and rear ones running full span, the one in the centre being of two-thirds span. Top and bottom skins are machined sheets, jointless from root to tip, with Redux-bonded stiffeners. The areas between spars are sealed to form integral tanks that carry most of the fuel. There is also a ventral tank and a small dorsal tank bringing total fuel capacity to 1,180 Imp/1,417 US gal (5,364 litres).

Double slotted flaps are moved with screw jacks that are driven by hydraulic motors. The pilot's control is gated at UP, 15 degrees (the take-off settings), 25 degrees (approach) and 45 degrees (landing). There is also a LIFT-DUMP setting (73 degrees) for use after touchdown. Ahead of the flaps are airbrakes that extend above and below the wing. The ailerons, elevators and rudder are mechanically operated via cables, and to lighten control loads they are fitted with geared tabs in addition to those used for trimming. To limit spanwise flow, wing fences are positioned about half way between the wing-tips and the fuselage sides. Sweep-back is 20 degrees measured at quarter chord.

A long dorsal fin commences from above the rear cabin window and blends with the very tall main fin, which continues down below the tail cone where it is joined by a small ventral area. Two-thirds up the fin is the swept tailplane and its separate elevators. Rather than fit a powered rudder (and at cruising speeds it is immovable with one foot) the 125 has always had an automatic rudder bias that takes the form of a double-sided piston. In normal flight it is maintained at an even pressure by bleed air from each engine. When a motor fails, pressure is lost on one side of the piston and it then moves, applying rudder in a corrective direction to counteract yaw. In the 700 series

the system is duplicated, but a little out-of-balance force remains following a power failure, otherwise pilots, finding it difficult to identify the failed engine, might shut down the wrong one and make things even quieter!

The fuselage, which is pressurized to 8.35 psi and can provide a sea-level cabin at aircraft altitudes of 21,500 ft (8,000-ft cabin at 41,000 ft) is very robustly constructed. In most small aircraft the wing passes through the fuselage, but in the 125 the fuselage sits on the wing, an arrangement which, from certain angles, makes the aircraft look somewhat pregnant. However, two important advantages that result from sacrificing a little of the aesthetics are first, the provision of a flat floor giving excellent headroom throughout the cabin area, and second, the structural charms of having a circular, pressure-tight fuselage that is unbroken by the intrusion of wing spars. Furthermore, all fuel lines, which are stainless steel, run outside the cabin area.

Six windows are provided on each side of the cabin, there are two large windows adjacent to each pilot and the windscreen takes the form of two electrically heated (for ice protection) flat screens set at an angle. Double

The most successful of the medium-size business jets, constructed to full airline standards.

The passenger cabin offers stand-up headroom from one end to the other, and outstanding levels of quietness and comfort.

wheels are fitted to all undercarriage legs and the hydraulically powered steering has a maximum turning angle of 45 degrees left and right of centre, a simple disconnecting pin allowing complete freedom through 360 degrees while towing. Hydraulic anti-skid (Maxaret) brakes of the multi-disc type use sintered iron pads.

Flying surface leading edges are given ice protection by a TKS fluid system, the windshields have gold film heating and the engines do their own thing with self-generated hot air. Electric power is supplied by the engine starter/generators which provide 12 kw. Full air-conditioning is provided and to sustain this while on the ground an AiResearch auxiliary power unit (APU) is fitted in the rear fuselage. This is also capable of generating electricity, thus making the BAe 125/700 completely self-sufficient at any airfield, however primitive.

The two Garrett fanjets are mounted in pods, with their air intakes protected by the wings from flying stones. They have very advanced, computerized fuel control and their low fuel consumption coupled with remarkable quietness have given the 125 series a massive shot in the arm.

A large door drops down from the left-hand side of the fuselage ahead of the wing. It has four steps and a rigid handrail. Viewed on the ground the BAe 125/700 stands out as a magnificent example of top-quality aeronautical engineering. It is rather larger than the Falcon 10 (see previous airtest) but more compact than a Falcon 20. From whatever angle you look at it, the BAe 125/700 is strikingly handsome.

Cabin and flight deck
The entrance door is 4 ft, 3 in (1.30 m) high and 2 ft, 3 in (69 cm) wide. You walk up the stairs to be confronted by a vestibule. Directly opposite is a 30 cu ft (0.85 cu m) baggage area (if required, an oven taking standard airline cooking trays may be installed nearby). On the right of the door is a refreshment unit with a small sink, and next to this a coat wardrobe. To the left is a door leading to the flight deck, to the right another one opens into the

cabin – and what a cabin! From front to rear there is an unobstructed 5-ft, 9-in (1.75-m) headroom and almost 6-ft width. Advanced concealed lighting has been developed to avoid shadows, there are automatic emergency oxygen masks that drop from the ceiling in the event of lost cabin pressure, as well as all the usual reading lights, tables and fresh-air vents. Up to fourteen passengers may be carried but the standard layout comprises a large three-seat settee down the left wall, four swivelling, tracking and reclining armchairs of a kind you will never see in an airliner, and a sideways-facing corner seat. Each position has its own table. To give some idea of the amount of space for passengers take a look at these comparisons:

Aircraft	Cu ft/passenger
Lear Jet 36	38
Boeing 727-200	41
Westwind 1124	47
Boeing 747	63
BAe 125/700	76

At the rear of the cabin is another door leading to an airliner-size toilet with an electrically-flushing WC, a simulated marble washbasin with hot and cold running water emerging from gold-plated taps (faucets in the US), a vanity unit and an electric shaver point. Follow that!

Up front you go through a door to enter a flight deck that measures up to the highest airline standards. There is a small control unit for the APU mounted on the door frame. In the roof is a large panel for the electrics, anti-icing, fuel system environmental control (cabin temperature, air-conditioning and so forth) and engine starting. There is a Collins FCS 80 autopilot/flight system, offering such facilities as Mach hold, coupled approach cleared to a decision height of only 100 ft, navigation modes, height lock, rate of climb/descent lock and even a 'half-bank' mode for when the chairman and his co-directors are suffering from brewer's twitch. Following big plane practice the autopilot mode controls are mounted in the glareshield.

All circuit breakers are on panels behind the copilot, who also has the simple pressurization controls below his flight panel. The usual engine instruments associated with Garrett fanjets are in the centre of the panel with the weather radar to their left and the avionics, directly below. These can include Omega/VLF long-range area navigation if required.

A wide console carries the two thrust levers, the flap/lift-dump lever, the parking brake and the three trim wheels. Nearby is a large 'T'-shaped handle which works the airbrakes. The pilots' controls take the form of ramshorns, which I find more comfortable to use than any other type. They carry buttons for electric trim, transmit/intercom between pilots or for cabin announcement, autopilot cut-out, a remote transponder ident and one that places the flight director 'V' bars in line with the aircraft symbol when flying manually. On the left ledge, set at an ideal angle for the captain, is the nosewheel steering knob.

I find it hard to fault this magnificent flight deck, except for a pair of rather crude tinted sun visors which have to be clipped into place. Like all else on this technological marvel everything in the flight deck is of the highest quality.

In the air

A little switch makes the APU burst into life and then there is current for starting without needing to use the battery. The right-hand engine starter was pressed, at 10 per cent N_2 (gas generator) rpm the high-pressure cock was opened, the ITT went up to 572°C and then the engine spooled up to allow selection of 'generator on line'. With both engines running I released the brakes and moved forward. Turns of only 30-ft radius can be made on the excellent nosewheel steering, which is matched by equally outstanding brakes.

For this flight we weighed 19,500 lb (8,845 kg) representing a typical load for two crew and eight passengers on a 700-nm trip with full IFR reserves. V_1/V_r was 109 kt at this weight and V_2 only 114 kt, which is lower than for some piston twins. With power partly set I released the brakes steering with the nosewheel knob, adding take-off power as we accelerated towards 70 kt, at which point the rudder became effective and I could move my hand to the control yoke. Except for the thump of wheels on joints in the runway surface there was utter silence on the flight deck as we hit rotate speed. I raised the nose to 12 degrees on the ADI, rapidly went through V_2 as the wheels came up and then settled at the 150-kt initial climbing speed. The VSI showed 3,500 ft/min. Ten minutes later we were at 28,000 ft going up at more than 2,300 ft/min.

At 29,000 ft I set up the high-speed cruise using 97.6 per cent N_1 (fan) rpm and 825 degrees ITT. Fuel flow was 1,960 lb (889 kg)/hr, or 245 Imp/290 US gal (1,098 litres) per hour, and our TAS was 460 kt. Next I climbed to 39,000 ft and had a look at the long-range cruise. You get 390–400 kt, according to conditions for a fuel burn of 1,200 lb (544 kg)/hr. The autopilot is outstanding and manual handling can only be described as exceptional, with very high stability to match.

They fit a stick shaker which threatens to wreck your watch when you approach the stall. With full-flap the 'g' break came, wings level, at only 80 kt. The airbrakes will provoke normal descent rates of around 10,000 ft/min.

Single-engine service ceiling is 25,000 ft, when the one-engine cruising speed is 250–280 kt according to weight. Engine-out climb is a reassuring 1,350 ft/min at average weight. At one stage I went back into the cabin and it is quieter than many of the big passenger jets, particularly in the forward seats.

By the time we were due to return, Hatfield were reporting a 900-ft ceiling and 2-km visibility – conditions that provided an opportunity for me to try some instrument flying on the 125. Initial approach was flown at a modest 117 kt, going for 107 kt on short finals. The aircraft went down the glideslope like a train pulling into the station, brushing aside turbulence like an ice-breaker. During the round-out the thrust levers were brought back, the main-wheels touched, I lowered the nosewheel, then applied lift-dump as the powerful brakes were eased on. They are a lucky breed, the pilots who fly the 125/700 on a day-to-day basis.

Capabilities

Out of a ramp weight of 25,000 lb (11,567 kg) – about the same as a Dakota – up to 9,440 lb (4,282 kg) can be fuel. Operating weight, which includes a 600-lb (272-kg) radio installation, 70 lb (32 kg) for refreshments and water, and two crew, is quoted at 13,885 lb (6,298 kg). Zero fuel weight is 16,050 lb (7,280 kg) and that is enough for ten people with their baggage – or fourteen folk of average weight 154 lb (70 kg) each. With the zero fuel weight of 16,050 lb (7,280 kg) added to 9,440 lb (4,282 kg) of maximum fuel we reach a total of 25,490 lb (11,562 kg), so it would seem that even a full cabin and full fuel can only get to within 10 lb (5 kg) of the maximum authorized weight. However, the BAe 125/700 is a full-tanks/full-cabin aircraft and the only consideration that might limit take-off weight would be the length of runway available at the points of departure and arrival. Under standard conditions the aircraft has a take-off distance of about 5,500 ft at maximum weight but it can take off in 4,000 ft with sufficient fuel for 1,500 nm. This level of field performance allows the aircraft to use airfields that are too small for the airliners.

Taking the full-fuel case the 125/700 would fly its full cabin for a with-reserve 2,660 nm at 395–400 kt while cruising at 35–41,000 ft. Alternatively, the high-speed power setting could fly you for 2,150 nm at 435 kt. Since the BAe 125/700 first appeared experience has shown that the Garrett engines are more economical than expected and the published maximum range has increased by 140 nm. The BAe 125/700 is a big aircraft for eight passengers, consequently its seat miles per unit of fuel are not as attractive as, for example, the much smaller and slower Citation II described earlier in the chapter. However, when you carry twelve passengers (maximum is fourteen), seat nautical miles per Imperial gallon are rather better at almost 23. Figures for US gallons and litres are 19.1 and 5.04, respectively. I have quoted seat miles per unit of fuel in a number of tests but this is only one facet of operating economy. There are others of equal importance.

Early models were weight limited at some airfields where average temperatures are high since this affects climb performance in the event of engine failure. Now the engine manufacturers have provided 'Automatic performance reserve' which ensures additional power under emergency conditions.

Verdict

The marketing boys at Hatfield are fond of saying that if you want a bizjet offering one inch more headroom and one knot more cruising speed it will cost another million dollars. The nearest competitor to the BAe 125/700 is the very beautiful Dassault-Breguet Falcon 20. It is larger, more powerful, has less payload range, offers no more internal space (if as much) and is no faster at recommended cruise.

To this day the 125 remains the smallest business jet with a true, stand-up/walk-around cabin. Of course an aircraft built to the specification of a 125/700 cannot be cheap. But it is an investment when applied to the right

Facts and figures

Dimensions

Wing span	47 ft
Wing area	353 sq ft
Length	50 ft, 8 in
Height	17 ft, 7 in
Cabin height	5 ft, 9 in
Width	5 ft, 11 in
Length (ex flight deck)	21 ft, 6 in

Weights & loadings

Max ramp/take-off	25,500 lb (11,567 kg)
Max landing	22,000 lb (9,979 kg)
Max zero fuel	16,050 lb (7,280 kg)
Operating weight (including two pilots)	13,885 lb (6,298 kg)
Useful load	11,615 lb (5,269 kg)
Seating capacity (with crew)	10/16
Max baggage	500 lb (227 kg)
Max fuel	1,180 Imp/1,417 US gal (5,363 litres)
Wing loading	72.24 lb/sq ft

Performance

High-speed cruise at 31,000 ft	435 kt
Long-range cruise at 35,000–41,000 ft	395 kt
Maximum range at 435 kt (31,000 ft)	2,150 nm
Maximum range at 395 kt (35,000-41,000 ft)	2,660 nm
Rate of climb Two engines	3,350 ft/min
One engine	1,350 ft/min
Take-off distance over 50 ft	5,500 ft
Landing distance over 50 ft	4,500 ft
Service ceiling Two engines (max operating)	41,000 ft
One engine	25,000 ft

Engines

2 × Garrett TFE 731-3R-1H fanjets giving 3,700 lb static thrust.

TBO

Not limited. Engines maintained on a contractual 'power-by-the-hour' basis.

task. This is best illustrated by the 'retrofit' business that has developed around the 125. Companies owning twelve-, fifteen- , and eighteen-year-old models that have long since been written off the books are spending £1 million or so on having the old engines replaced by Garrett turbofans, an interior face-lift and a flight deck update. The truth of the matter is that the airframes do not wear out.

Although the 125/700 is one of the more expensive busi-ness jets on the market more than 560 had been sold by July 1983. What a pity they never called it the Jet Dragon. This wonderful aircraft deserves an appropriate name.

In the air the 125/700 is every inch the scaled-down airliner. Engineering standards are matched by few other bizjets.

HOW DO THEY RATE?

The following assessments compare aircraft of similar class. For example, four stars (Above average) means when rated against other designs of the same type.

★★★★★ Exceptional
★★★★ Above average
★★★ Average
★★ Below average
★ Unacceptable

Aircraft type	Appearance	Engineering	Comfort	Noise level	Visibility	Handling	Payload/ Range	Field Performance	Cruising Speed	Economy	Single-engine Climb	Value for Money
Cessna 550 Citation II	★★★	★★★	★★★	★★★	★★★★	★★★	★★★	★★★★	★★★	★★★★	★★★	★★★★
Dassault-Breguet Mystère Falcon 10	★★★★★	★★★★★	★★★★	★★★★	★★★★	★★★★★	★★★★	★★★★	★★★★★	★★★★	★★★★★	★★★★
British Aerospace 125/700	★★★★★	★★★★★	★★★★★	★★★★★	★★★★★	★★★★★	★★★★★	★★★★	★★★★	★★★★	★★★	★★★★★